COCHLEAR NONLINEARITY

A computational model of the cochlea solved in the frequency
domain.

Eerste druk, mei 1995

Tweede druk, september 2009

© 1995 Luc Johan Kanis

Illustratie boekomslag Luc Johan Kanis

ISBN: 978-1-4092-9403-0

NUR: 910

COCHLEAR NONLINEARITY

A computational model of the cochlea solved in the frequency
domain

ACADEMISCH PROEFSCHRIFT

ter verkrijging van de graad van doctor
aan de Universiteit van Amsterdam
op gezag van de Rector Magnificus
prof. dr. P.W.M. de Meijer
ten overstaan van een door het college van dekanen ingestelde
commissie in het openbaar te verdedigen in de Aula der Universiteit
op vrijdag 19 mei 1995 te 14.00 uur door

Lucas Johannes Kanis

geboren te Amsterdam

PROMOTIECOMMISSIE

Promotor:

Prof. Dr. E. de Boer (Faculteit der Geneeskunde, UvA)

Overige leden:

Prof. Dr. P.J. Schouwenburg (UvA)
Prof. Dr. Ir. W.A. Dreschler (UvA)
Prof. Dr. Ir. H. Spekreijse (UvA)
Prof. Dr. Ir. H. Duifhuis (RUG)
Prof. Dr. Ir. H.P. Wit (RUG)
Prof. Dr. G.F. Smoorenburg (RUU/TNO-ITM)
Dr. V.F. Prijs (RUL)

This work has been supported by the Netherlands Foundation of Pure Research (NWO)

Contents

1

Introduction

The ear is a highly specialized organ that converts sound signals into neural impulses which carry the auditory information to the brain. The ear can be divided into three parts: the *outer ear*, the *middle ear* and the *inner ear* (*cochlea* and *semicircular canals*), see Fig. 1.

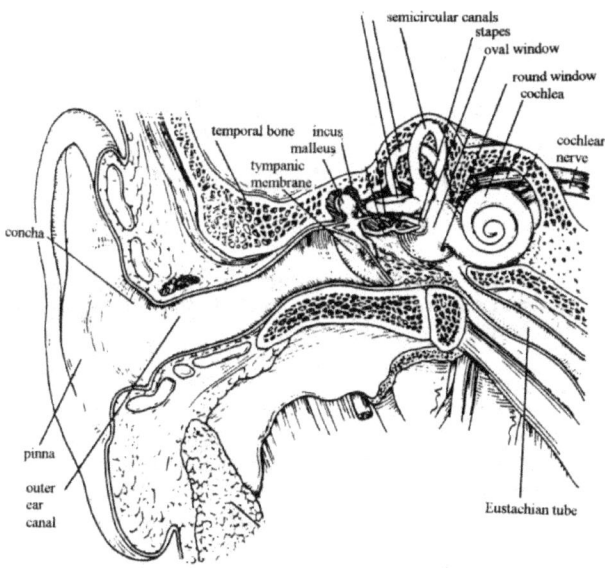

Figure 1. The external, middle, and inner ears in man. From *Tissues and Organs: A Text-Atlas of Scanning Electron Microscopy*, by R.G. Kessel and R.H. Kardon (W.H. Freeman and Company).

The outer ear picks up a sound wave and transfers it through the *outer ear canal* to the *eardrum*. This is done in such a way that the sound pressure at the eardrum is amplified in the range of frequencies that are important for speech recognition. By changing the spectral density of the sound the outer ear provides some cues which assist us in localizing sound. The middle ear consists of a chain of three ossicles, called the *malleus, incus* and *stapes*, that transmits the sound vibrations of the *tympanic membrane* (ear drum) to the *oval window*, the entrance to the cochlea. During this transfer the pressure is amplified by a factor of approximately 20 which reduces the reflection occurring when a wave is transmitted from a low-density medium such as air to the high-density cochlear fluid. Another property of the middle ear is to protect the cochlea against overloading. This mechanism comes into operation not only when loud sounds enter the ear but also during one's own speech (even before that speech is initiated).

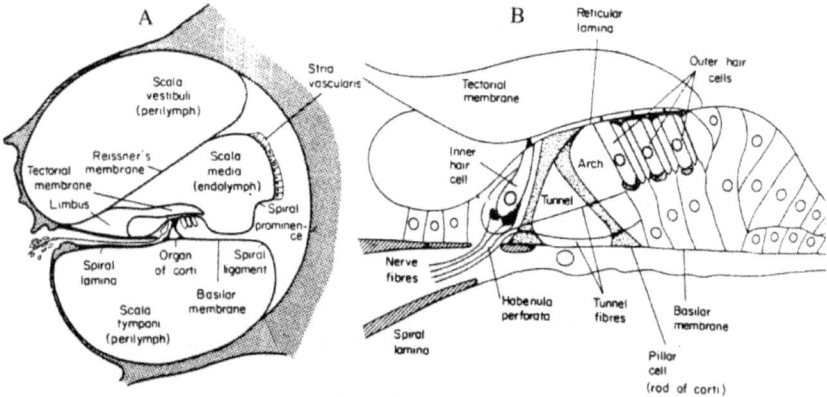

Figure 2. (a) A cross-section of the cochlear duct. (b) A cross-section of the organ of Corti, as it appears in the basal turn. From *An introduction to the Physiology of Hearing*, by J.O. Pickles (Academic Press, San Diego).

As the oval window is moved (in an oscillatory way) by the stapes, a sound wave arises in the fluid contained in the spiralling tunnel of the cochlea. The associated sound pressure causes a partition consisting of the *basilar membrane* and the *organ of Corti* (where the sensory cells are located, see Fig. 2) to move. As a result of the interaction between fluid and partition a travelling wave is set up on the basilar membrane as well as in the fluid. In the case where the stimulus is sinusoidal the amplitude of the wave increases along the length of the partition until it reaches a maximum at a location that depends on the frequency. Higher frequencies are mapped at locations near the stapes, while low-frequency tones arrive at places further inside the cochlea. The reason for this frequency mapping is that the stiffness that can be attributed to the cochlear partition decreases from stapes to *helicotrema* (i.e., the other end of the cochlea). When the stimulus consists of more than one component, every component gives rise to a travelling wave that peaks at a different location. The cochlea thus performs some kind of

8

frequency analysis of the auditory stimulus.

In the organ of Corti two types of sensory cells are present, the *inner hair cells* and the *outer hair cells*, see Fig. 2(b). During movements of the basilar membrane a shearing motion between the underside of the *tectorial membrane* and the tops of these hair cells arises. This leads to excursions of the *stereocilia* of the hair cells and subsequent polarization of the cells. In the case of inner hair cells, polarization leads to the creation of action potentials in the *auditory nerve*. Polarization of the outer hair cells gives rise to cell length changes that are probably involved in improving sensitivity and frequency selectivity of the ear.

Before 1971 it was thought that the BM response was a purely linear function of input level. Then, Rhode (1971) showed that for single-tone stimulation the BM response behaves nonlinearly as a function of input level: the response is compressed for stronger stimuli. Rhode also showed that after death the response becomes linear. This implies that cochlear functioning depends on the physiological state of the cochlea. Not only is the amplitude of the response influenced by nonlinearity of the cochlea, but the phase is also influenced. Rhode and Robles (1974) and Sellick *et al.* (1982) reported an increase of phase lag with increasing stimulus level for frequencies below the characteristic frequency (this is the frequency for which the location under study is the most sensitive), and a decrease for frequencies above this frequency. Corresponding phase behaviour has been found in inner hair cell responses (e.g., Nuttall and Dolan, 1993) and in physiological responses (Anderson *et al.*, 1971).

Other nonlinear phenomena have also been discovered. During stimulation with two tones the response to one tone can be suppressed by the presence of the second tone. This phenomenon, called two-tone suppression, was first demonstrated in cochlear microphonic recordings (Covell and Black, 1936), but later also in neural recordings (Sachs and Kiang, 1968), inner hair cells (Sellick and Russel, 1979) and the basilar membrane (Rhode, 1977). During two-tone suppression the phase of the probe tone exhibits changes greatly dependent on stimulus conditions (Cheatham and Dallos, 1990; Nuttall and Dolan, 1993): not only the frequency of the suppressing tone but also its input level determines the character of the phase shifts.

Another nonlinear phenomenon that occurs when the cochlea is stimulated with two tones is the generation of combination tones. When humans listen to pairs of tones they can hear tones that are not present in the stimulus (Goldstein, 1967; Smoorenburg, 1972). Counterparts of these tones have been measured in neural responses (Goldstein and Kiang, 1968), in inner hair cells (Nuttal and Dolan, 1990), in otoacoustic emissions (Kemp, 1979) and also on the level of the basilar membrane (Robles *et al.*, 1991).

Although several nonlinear models of the cochlea have been devised in the past (a description of these models is given later in this section), the nonlinearities referred to above have not yet been fully understood and replicated. The main purpose of this thesis is to devise a cochlea model that replicates the nonlinear phenomena better than previous models and to solve that model in such a way that insight into the mechanisms of nonlinearity is more easily acquired than with existing solution methods.

First, a linear version of the model should be able to reproduce BM motion measured at low levels where the cochlea operates in its linear regime. This touches the problem of 'activity' in the cochlea. Actually, the older BM responses measured by, e.g., Rhode (1971) and Johnstone and Yates (1974) can be simulated very well by a passive model (Viergever and Diependaal, 1983), i.e., a model in which no active elements are present. However, it is harder to match the newer data showing much sharper tuning at low stimulus levels (Sellick et al., 1982; Robles et al., 1986) with such a model (Viergever and Diependaal, 1986). As a way out of this dilemma the concept of local activity - at certain locations along the length of the cochlea the cochlear partition injects more energy into the system than it absorbs - was introduced by Kim et al. (1980). Later, a more refined, micromechanical, model containing a secondary resonance of the TM, was formulated by Neely (1981) and Neely and Kim (1983, 1986). In the latter paper the outer hair cells were modelled as pressure sources controlled by relative movements of the tectorial membrane and the reticular lamina. This was consistent with the *in vitro* observation that a hair cell is polarized when its hair bundle is deflected (Hudspeth and Corey, 1977) and that in an outer hair cell this polarization gives rise to length changes (Brownell et al., 1985). The idea of local activity was supported by theoretical evidence (de Boer, 1983a, 1983b) that, in the class of models considered here, the sharp BM responses as measured by Sellick et al. (1982) could only be obtained if the resistance part of the BM impedance was negative over a limited region of the cochlear partition. As a critical note it should be mentioned that in the models by Kolston (1988), Kolston and Viergever (1989), and Kolston et al. (1989) frequency selectivity is obtained by changing the stiffness of the partition, but this is at the expense of a reduced sensitivity.

Soon after 1971 several nonlinear cochlea models were developed that tried to replicate Rhode's findings (1971) on nonlinear behaviour of the BM response. In these models a nonlinear term that made the resistance increase with increasing response was inserted in the partition impedance (Hubbard and Geisler, 1972; Kim et al., 1973; Hall, 1974). The model by Hall predicted in the case of two-tone stimulation the generation of distortion products and their presence at locations where the primaries were not detectable. This prediction was verified by Kim et al. (1980) at the neural level, and at the mechanical level by Robles et al. (1991). Nonlinear cochlea models by Kim et al. (1973) and Hall (1977) and Zwicker's hardware model (1979, 1986) were able to simulate two-tone suppression with suppressors with higher frequencies than the probe frequency but not with lower frequencies. Only by postulating a second filter between the mechanical response and neural excitation signal could Hall solve the discrepancy between experimental and model results. In 1992 Ruggero et al. observed low-side two-tone suppression in the BM response, so the need for a model that could explain this phenomenon on the mechanical level was felt again. Cohen and Furst (1993) gave an example of mutual suppression in a model with an *ad hoc* activity distribution.

In this thesis we describe a locally-active cochlea model in which active elements generate pressures that are added directly over the basilar membrane similarly as in the model of Neely and Kim (1986). Our model is even simpler in that the influence of the tectorial membrane and stereocilia on the mechanics of the cochlear partition is left out of consideration. In line with experimental findings (cf. Patuzzi et al., 1989) we made the pressure generated by the OHCs a nonlinear function of its input. Thus all nonlinearity is found in the generation of active pressures by the OHCs. At the time that this model was developed two similar models were devised by Geisler et al. (1993) and Neely and Stover (1993) that were solved in the time domain.

Instead of solving our model in the time domain, which has been the standard procedure for nonlinear models, we chose to solve the model in the frequency domain. We only consider the amplitude and phase of the system variables in steady state and not the detailed time-evolution of the variables. This has the advantage that the concept of impedance can be used which enables us to monitor the effect of response changes on the impedance and vice versa. Also, we have used the concept of characteristic impedance to decrease the amount of reflection at the stapes. The quasilinear method, as we have called it, splits the system variables into their Fourier components resulting in discrete sets of equations belonging to these Fourier components. Thus, although the different responses are not independent (they may suppress each other) we may solve them separately. Separation is possible if, as in our cochlea model, at *every* location the basilar membrane is *entrained* by the stimulus, i.e., the basilar membrane oscillates periodically with period frequency equal to the stimulus period frequency.

In chapter 2 the cochlea model is solved with the quasilinear method under single-tone stimulation. It is shown that the frequency selectivity and the sensitivity of the velocity response of the basilar membrane are greatly enhanced by the presence of pressure sources inside the cochlea if these have the right place-frequency dependence. This dependence is such that in a limited region basal to the response peak the resistance component of the BM impedance is negative. Because we have represented the pressure source as a saturating function of its input the response is a nonlinear function of the intensity of the sound stimulus. Increasing stimulus intensity has a 'flattening' effect on the response. That is, the sharpness of the response peak decreases, and thereby the sensitivity and frequency selectivity of the ear. The flattening happens because the response brings the pressure generation into saturation. As a consequence the response is amplified less. We have called this process *self-suppression*. At high intensities the response becomes similar to the linear passive response, so that the input-output function is found to be linear for input levels below 20 and above 90 dB SPL with a nonlinear transition in between. This is consistent with results of mechanical experiments (Sellick *et al.*, 1982; Robles *et al.*, 1986).

Chapter 3 discusses the phenomenon of two-tone suppression. During this phenomenon the amplitude of a single-tone response may decrease if a secondary tone is added to the sinusoidal stimulus. Also the phase may change. Distinction is made between dominant and non-dominant suppression where dominant (non-dominant) means that the suppressor has a larger (smaller) velocity amplitude than the probe at its peak location. In experiments by Nuttall and Dolan (1993) it was found that the sign of the probe's phase change depends on which of these two conditions is chosen. This led Nuttall and Dolan to suggest that different suppression mechanisms exist. We show that there is no need for such a conjecture. Furthermore, the attenuation hypothesis which states that the addition of a suppressor has the same effect on the probe as probe level attenuation, is rejected in this chapter. Cheatham and Dallos (1990) had already pointed out that, under certain stimulus conditions, a mismatch exists between predictions made by this hypothesis and experimental findings, but they did not reject the hypothesis.

In chapter 4 the experimental finding that distortion products present in evoked cochlear emissions are 'tuned' as a function of frequency is replicated with the model. It has been argued by several authors (Brown and Gaskill, 1990; Brown and Williams, 1993; Allen and Fahey, 1993) that this 'tuning' is the consequence of a filtering mechanism in the ear: distortion products in the pressure generated by the outer hair cells are supposed to be filtered as they are coupled back to the BM. However, in our model the pressures are not filtered after they have been generated and still similar 'tuning' is observed in the model. It is shown that the 'tuning' is a nonlinear effect since it disappears

11

at low input levels. Furthermore, a prediction is given about the response of the $2f_1-f_2$ combination tone at its peak location as a function of primary frequency ratio.

In chapter 5 a replication of the experiment performed by Allen and Fahey (1992) is discussed. While keeping both the frequency of the $2f_1-f_2$ combination tone (CDT) and the physiological response at the location characteristic for the CDT constant they measured the CDT emission as a function of primary frequencies. No significant change in emission was found. They concluded that the cochlea does not possess an amplifying mechanism. This conclusion is challenged in this chapter. Similar results as those obtained by Allen and Fahey were achieved with our cochlea model. It is therefore concluded that Allen and Fahey's interpretation of their experiment is incorrect. Careful analysis shows that they compared two stimulus conditions that were quite similar and therefore could not be expected to give a large difference in emission. It is furthermore explained that the relation between difference in emission and the gain of the amplification process is much more complicated than they supposed.

Chapter 6 compares time-domain solutions with solutions obtained with the quasilinear method. This is done for single-tone as well as for two-tone stimulation. It is concluded that, as long as the primary components are concerned, stable nonlinear cochlea models can be solved perfectly in the frequency domain. For combination tones the agreement between quasilinear and time-domain solutions is less perfect but still satisfactory.

References

Allen, J.B. and Fahey, P.F. (1992). "Using acoustic distortion products to measure the cochlear amplifier gain on the basilar membrane," J. Acoust. Soc. Am. 96, 178-188.

Allen, J.B., and Fahey, P.F. (1993). "Evidence for a second cochlear map," in: *Biophysics of Hair Cell Sensory Systems*, edited by H. Duifhuis, J.W. Horst, P. van Dijk, and S.M. van Netten (World Scientific, Singapore), pp. 296-302.

Anderson, D.J., Rose, J.E., Hind, J.E., and Brugge, J.F. (1971). "Temporal position of discharge in single auditory nerve fibers within the cycle of a sine-wave stimulus: Frequency and intensity effects, " J. Acoust. Soc. Am. 49, 1131-1139.

Boer, E. de (1983a). "No sharpening? A challenge for cochlear mechanics," J. Acoust. Soc. Am. 73, 567-573.

Boer, E. de (1983b). "On active and passive cochlear models - towards a generalized treatment," J. Acoust. Soc. Am. 73, 574-576.

Brown, A.M., and Gaskill, S.A. (1990). "Can basilar membrane tuning be inferred from distortion measurement?," in *Mechanics and Biophysics of Hearing*, edited by P. Dallos, C.D. Geisler, J.W. Matthews, M.A. Ruggero, and C.R. Steele (Springer-Verlag, Berlin), pp. 164-169.

Brown, A.M., and Williams, M.W. (1993). "A second filter in the cochlea," in: *Biophysics of Hair Cell Sensory Systems*, edited by H. Duifhuis, J.W. Horst, P. van Dijk, and S.M. van Netten (World Scientific, Singapore), pp. 72-77.

Brownell, W.E., Bader, C.R., Bertrand, D., and Ribaupierre, Y. de (1985). "Evoked mechanical responses of isolated cochlear outer hair cells," Science 227, 194-196.

Cheatham, M.A., and Dallos, P. (1990). "Comparison of low- and highside two-tone suppression in inner hair cell and organ of Corti responses," Hear. Res. 50, 193-210.

Cohen, A., and Furst, M. (1993). "Cochlear model for rate suppression based on cochlear amplifier dynamics," in: in *Biophysics of Hair-cell Systems*, edited by H. Duifhuis, J.W. Horst, P. van Dijk and S.M. van Netten (World Scientific, Singapore), pp. 323-329.

Covell, W.P., and Black, L.J. (1936). "The cochlear response as an index of hearing," Am. J. Physiol. 116, 524-530.

Geisler, C.D., Bendre, A., and Liotopoulos, F.K. (1993). "Time-domain modeling of a nonlinear, active model of the cochlea," in *Biophysics of Hair-cell Systems*, edited by H. Duifhuis, J.W. Horst, P. van Dijk and S.M. van Netten (World Scientific, Singapore), pp. 330-337.

Goldstein, J.L. (1967). "Auditory nonlinearity," J. Acoust. Soc. Am. 41, 676-689.

Goldstein, J.L. and Kiang, N.Y.S. (1968). "Neural correlates of the aural combination tone 2f1-f2," Proc. IEEE 56, 981-992.

Hall, J.L. (1974). "Two-tone distortion products in a nonlinear model of the basilar membrane," J. Acoust. Soc. Am. 56, 1818-1828.

Hall, J.L. (1977). "Two-tone suppression in a nonlinear model of the basilar membrane," J. Acoust. Soc. Am. 61, 802-810.

Hubbard, A.E., and Geisler, C.D. (1972). "A hybrid-computer model of the cochlear partition," J. Acoust. Soc. Am. 51, 1895-1903.

Hudspeth, A.J., and Corey, D.P. (1977). "Sensitivity, polarity, and conductance change in the response of vertebrate hair cells to controlled mechanical stimuli," Proc. Natl. Acad. Sci. USA 74, 2407-2411.

Johnstone, B.M., and Yates, G.K. (1974). "Basilar membrane tuning curves in the guinea pig," J. Acoust. Soc. Am. 55, 584-587.

Kemp, D.T. (1979). "Evidence of mechanical nonlinearity and frequency selective wave amplification in the cochlea," Arch. Otorhinolaryngol. 224, 37-45.

Kim, D.O., Molnar, C.E. and Pfeiffer, R.R. (1973). "A system of nonlinear differential equations modeling basilar-membrane motion," J. Acoust. Soc. Am. 54, 1517-1529.

Kim, D.O., Molnar, C.E. and Matthews, J.W. (1980). "Cochlear mechanics: Nonlinear behavior in two-tone responses as reflected in cochlear-nerve-fiber responses and in ear-canal sound pressure," J. Acoust. Soc. Am. 67, 1704-1721.

Kolston, P.J. (1988). "Sharp mechanical tuning in a micromechanical cochlear model," J. Acoust. Soc. Am. 83, 1481-1486.

Kolston, P.J. and Viergever, M.A. (1989). "Realistic basilar membrane tuning does not require active elements," in: Cochlear Mechanisms, Structure, Function and Models, edited by J.P. Wilson and D.T. Kemp (Plenum Press, London), pp. 415-424.

Kolston, P.J., Viergever, M.A., de Boer, E., and Diependaal, R.J. (1989). "Realistic mechanical tuning in a micromechanical cochlear model, J. Acoust. Soc. Am. 86, 133-140.

Neely, S.T. (1981). "Fourth-order partition dynamics for a two-dimensional model of the cochlea," doctoral dissertation (Washington University, St. Louis).

Neely, S.T. and Kim, D.O. (1983). "An active cochlear model showing sharp tuning and high sensitivity," Hear. Res. 9, 123-130.

Neely, S.T. and Kim, D.O. (1986). "A model for active elements in cochlear biomechanics," J. Acoust. Soc. Am. 79, 1472-1480.

Neely, S.T., and Stover, L.J. (1993). "Otoacoustic emissions from a nonlinear, active model of cochlear mechanics," in: Biophysics of Hair Cell Sensory Systems, edited by H. Duifhuis, J.W. Horst, P. van Dijk, and S.M. van Netten (World Scientific, Singapore), pp. 64-70.

Nuttall, A.L., and Dolan, D.F. (1990). "Inner hair cell responses to the $2f_1$-f_2 intermodulation distortion product," J. Acoust. Soc. Am. 87, 782-790.

Nuttall, A.L., and Dolan, D.F. (1993). "Two-tone suppression of inner hair cell and basilar membrane responses in the guinea pig," J. Acoust. Soc. Am. 93, 390-400.

Patuzzi, R.B., Yates, G. K. and Johnstone, B.M. (1989). "Outer hair cell receptor current and sensorineural hearing loss," Hear. Res. 42, 47-72.

Rhode, W.S. (1971). "Observations of the vibration of the basilar membrane in squirrel monkeys using the Mössbauer technique," J. Acoust. Soc. Am. 49, 1218-1231.

Rhode, W.S. (1977). "Some observations on two-tone interaction measured with the Mössbauer effect," in: Psychophysics and Physiology of Hearing," edited by E.F. Evans and J.P. Wilson (Academic Press, London), pp. 27-38.

Rhode, W.S., and Robles, L. (1974). "Evidence from Mössbauer experiments for nonlinear vibration in the cochlea," J. Acoust. Soc. Am. 55, 588-596.

Robles, L., Ruggero, M.A. and Rich, N.C. (1986). "Basilar membrane mechanics at the base of the chinchilla cochlea. I. Input-output functions, tuning curves, and response phases," J. Acoust. Soc. Am. 80, 1364-1374.

Robles, L., Ruggero, M.A. and Rich, N.C. (1991). "Two-tone distortion in the basilar membrane of the cochlea," Nature 349, 413-414.

Ruggero, M.A., Robles, L., and Rich, N.C. (1992). "Two-tone suppression in the basilar membrane of the cochlea: Mechanical basis of auditory-nerve rate suppression," J. Neurophysiol. 68, 1087-1099.

Sachs, M.B., and Kiang, N.Y.-S. (1968). "Two-tone inhibition in auditory-nerve fibers," J. Acoust. Soc. Am. 43, 1120-1128.

Sellick, P.M., Patuzzi, R. and Johnstone, B.M. (1982). "Measurement of basilar membrane motion in the guinea pig using the Mössbauer technique," J. Acoust. Soc. Am. 72, 131-141.

Sellick, P.M., and Russell, I.J. (1979). "Two-tone suppression in cochlear hair cells," Hear. Res. 1, 227-236.

Smoorenburg, G.F. (1972). "Combination tones and their origin," J. Acoust. Soc. Am. 52, 615-632.

Viergever, M.A., and Diependaal, R.J. (1983). "Simultaneous amplitude and phase match of cochlear model calculations and basilar membrane vibration data," in Mechanics of Hearing, edited by E. de Boer and M.A. Viergever (University Press, Delft), pp. 53-61.

Zwicker, E. (1979). "A model describing nonlinearities in hearing by active processes with saturation at 40 dB," Biol. Cybernetics 35, 243-250.

Zwicker, E. (1986). "A hardware cochlear nonlinear preprocessing model with active feedback," J. Acoust. Soc. Am. 80, 146-153.

2

Self-suppression in a locally active nonlinear model of the cochlea: A quasilinear approach[*]

Abstract   Mechanical input-output functions of the cochlea for pure-tone stimuli are nonlinear for frequencies around the characteristic frequency. To simulate these functions, a long-wave model of the cochlea containing a saturating pressure generator (located at the site of the outer hair cells) is solved in the frequency domain with a quasi-linear method. In this method distortion products in the basilar-membrane (BM) response are treated as perturbations and the nonlinear pressure waveform is approximated by the first-order Fourier component. Because the suaturating pressure generator forms part of a feedback loop the solution of the model is achieved in a number of iteration steps. Model results show flattening of the BM response at higher input pressures; this property, called self-suppression, is due to saturation of the pressure generator. The resulting input-output functions display the main features of experimental curves. The third-order distortion product in the BM velocity is always more than 25 dB below the primary BM velocity and does not influence the results of the computation; this justifies the use of the quasi-linear method.

2.1 Introduction

In 1971 Rhode showed that the mechanical response of the basilar membrane (BM) to sinusoidal stimuli becomes less frequency-selective for stronger stimuli. This finding was extended in more refined experiments (Rhode, 1978; LePage and Johnstone, 1980; Sellick et al., 1982; Khanna and Leonard, 1982; Patuzzi et al., 1984a; Johnstone et al., 1986; Robles et al., 1986a; Nuttal et al., 1990; Ruggero and Rich, 1991). In the last two decades several models have been proposed to explain nonlinear phenomena in the cochlea (Hubbard and Geisler, 1972; Kim et al., 1973; Hall, 1974; Matthews, 1980; Duifhuis et al., 1985; Strube, 1985, 1986; Zwicker, 1986). Most of these nonlinear

---

[*] Preliminary results were presented during the International Meeting on *Auditory Processing of Complex Sounds* at the Royal Society in London, 1991.

models were solved in the time-domain - none of them in the frequency domain. For general reviews of this work on nonlinear models the reader is referred to Kim (1985) and de Boer (1991).

Our aim is to simulate cochlear nonlinearity in a *locally-active*[1] cochlear model. The model we develop follows earlier active models in that it is based on a BM impedance with a place-dependent active term that renders the BM resistance negative in a *restricted* region of the cochlea (Kim *et al.*, 1980a; de Boer, 1983, 1991; Neely and Kim, 1983, 1986; Geisler, 1991). As in the models of Neely and Kim we use a secondary resonance to achieve the proper activity distribution and a large cochlear gain. We make sure that the model is zero-point stable because in the healthy ear spontaneous emissions occur at only a few frequencies (Kemp, 1979; Zurek, 1981). In the present paper we restrict ourselves to input-output relations for sinusoidal stimuli.

We assume, more specifically, that in each cross-section of the cochlear channels outer hair cells (OHCs) generate a pressure on a cycle-to-cycle basis (Gitter and Zenner, 1988), that this pressure is added to the pressure difference across the Organ of Corti and that the BM velocity is modified accordingly. This means that the OHCs are the producers of local activity and form parts of a feedback system. Furthermore, it is assumed that the pressure-generating process is nonlinear and that this constitutes the *only* nonlinearity in the model. The saturating form of the nonlinearity has been inspired by physiological experiments (Hudspeth and Corey, 1977; Ashmore, 1987; Cody and Russel, 1987; Patuzzi *et al.*, 1989). Because of the saturation the model becomes less active, i.e., *de-activated*, when the stimulus is stronger. We will refer to the effect of de-activation on the BM response caused by increasing the stimulus input as self-suppression.

Because we would like to gain insight into the relation between de-activation and reduction of frequency selectivity in the BM response, we solve the nonlinear model with a *quasi-linear* method which operates in the frequency domain. Another reason to use this method is that we are mainly interested in global aspects of the BM response, and that time-domain computations, even for simple long-wave models, are very time-consuming.

A study of mechanical aspects of two-tone suppression (cf. Rhode, 1977; Robles *et al.*, 1986b; Ruggero *et al.*, 1992, Patuzzi *et al.*, 1984b) and combination tones (cf. Robles *et al.* 1990, 1991) is also possible with the proposed method. Results of these applications have been reported elsewhere (Kanis and de Boer, 1993).

2.2 The quasi-linear method

The basic assumption of the quasi-linear solution method is that distortion products do not noticeably influence the primary components of the BM velocity. We consider the model as driven by a sinusoidal signal and consider all system variables as split up into Fourier components. Then, because only sinusoidal signals are involved, a linear solution method can be used to find the variables. That the aforementioned assumption is justified is shown in Appendix B.

---

[1] Our definition of local activity: At certain locations in the cochlea more acoustical power is produced by the cochlear partition than absorbed by it.

18

The nonlinear pressure waveform that appears in our model is approximated by the appropriate Fourier component (for instance, the one corresponding to the primary frequency) at all locations in the cochlea. The BM velocity is solved by a linear method with the thus-obtained pressure distribution. Since in our model effects of activity involve feedback, we must iterate the solution. Parameters are 'updated' on the basis of the velocity distribution obtained in the preceding step, and used to solve for a new velocity distribution. The process is repeated until the distribution does not change noticeably between two subsequent steps. The resulting time gain with respect to time-domain computations is considerable. Details about the method are given in section II, part D.

2.3 Development of the method

2.3.1 The long-wave model

The cochlea is modelled as a straight fluid-filled narrow tube in the x-direction, equally divided into two rectangular scalae by a movable partition consisting of the basilar membrane (BM) and the Organ of Corti. The cochlea is driven at the stapes ($x = 0$) and short-circuited at the other end (at $x = x_{end}$) by means of a fluid connection between the two scalae, the helicotrema. We have assumed that all longitudinal coupling is through the fluid.

For simplicity a long-wave model is considered. The long-wave model accounts for all physical aspects of cochlear models in which the fluid dynamics is considered linear with the exception of the phenomenon of boundary-layer absorption (Lighthill, 1981), virtual mass of the channel fluid (de Boer, 1982) and the transition of long to short waves (de Boer, 1984, chapter 6). The one-dimensional fluid dynamics of such a linear long-wave model is described by the wave equation

$$p_{xx}(x;\omega) + k^2(x;\omega)\, p(x;\omega) = 0, \tag{2.1}$$

where $p(x;\omega)$ is the complex pressure in one of the scalae, and ω the radian frequency. The subscript xx denotes the second derivative with respect to x. The complex coefficient $k(x;\omega)$ has the dimension of a wavenumber, and it is defined by

$$k^2(x;\omega) = \frac{-2i\omega\rho}{hZ_{BM}(x;\omega)}, \tag{2.2}$$

in which ρ is the density of the fluid, h the effective height of the scalae, and $Z_{BM}(x;\omega)$ the impedance function that describes the BM. If $hZ_{BM}(x;\omega)$ were a purely negative imaginary constant, the coefficient $k(x;\omega)$ would be a real constant and denote the wavenumber of the solution: 2π divided by the wavelength. When $hZ_{BM}(x;\omega)$ varies slowly with x, $k(x;\omega)$ acquires the meaning of the local wavenumber (which might be complex).

One boundary condition needed to solve this cochlea model concerns the coupling to the middle ear and to the driving source. This is described in Appendix A. The other boundary condition is

$$p(x_{end}) = 0 \tag{2.3}$$

Once the wave equation has been solved, the BM velocity $v_{BM}(x;\omega)$ is calculated from

$$v_{BM}(x;\omega) = -2p(x;\omega) / Z_{BM}(x;\omega). \tag{2.4}$$

Implementation for a numerical solution is straightforward. The solution method is the same as the one used in de Boer (1980, Appendix A which is based on Allen, 1977). We use this method in every step of iteration in our quasi-linear solution. The program has been written in Turbo Pascal®. Since Turbo Pascal® does not support complex arithmetic, a 'unit' has been developed for this purpose[2]. In all examples the stimulus frequency is 6 kHz, so that the helicotrema can be set at 1 cm without loss of accuracy. The length of the cochlear partition is divided into 500 sections. Using fewer than 150 sections gives an 'impaired' response due to discretization errors.

1.3.2 The linear passive case

For the passive BM impedance function $Z^p_{BM}(x;\omega)$ we take the usual form

$$Z^p_{BM}(x;\omega) = i\omega M + R(x) + S(x) / i\omega, \tag{2.5}$$

with $S(x)$ and $R(x)$ given by

$$R(x) = \delta \sqrt{MS_0} \, \exp(-\alpha x / 2), \tag{2.6.a}$$

$$S(x) = S_0 \exp(-\alpha x). \tag{2.6.b}$$

The parameters are: $S_0 = 10^{10}$ [kg m^{-2} s^{-2}], $\alpha = 3 \times 10^2$ [m^{-1}], and $M = 0.5$ [kg m^{-2}]. These parameters represent the human cochlea (de Boer, 1980). The coefficient δ is the damping parameter and it is taken as constant: $\delta = 0.4$. The impedance $Z^p_{BM}(x;\omega)$ takes the place of $Z_{BM}(x;\omega)$ in Eqs. (2.2) and (2.4).

In Fig. 1 model results of the passive long-wave model are plotted as dashed lines. Part (a) shows the resistance component $R(x)$ given by Eq. (2.6.b) as the thick line; the reactance component is drawn as the thin line (we refer to the legend for details about

---

[2] In this unit a complex arithmetic function produces upon result a pointer to a complex number, where a complex number is defined as a record with two fields each containing a real number 6 bytes. More detailed information about the implementation of this unit can be obtained from the authors.

the scaling). In part (b) the magnitude of the BM velocity computed from Eq. (2.4) is plotted in dB relative to the stapes velocity. Note that in this, the passive, case the response peak is broad and not well pronounced. The solid lines and the arrows in this figure will be referred to later.

2.3.3 The linear locally-active case

Let us now introduce the outer hair cells (OHCs). We assume that they generate a pressure $P_{OHC}(x;\omega)$ that is added to the pressure difference across the BM. If we write this pressure as

$$P_{OHC}(x;\omega) = Z_{OHC}(x;\omega)\, v_{BM}(x;\omega), \tag{2.7}$$

the passive BM impedance $Z^p_{BM}(x;\omega)$ is modified by the feedback to the locally-active BM impedance $Z^a_{BM}(x;\omega)$ according to

$$Z^a_{BM}(x;\omega) = Z^p_{BM}(x;\omega) - Z_{OHC}(x;\omega). \tag{2.8}$$

One possible form of the impedance $Z_{OHC}(x;\omega)$ is obtained by considering a resonance of the tectorial membrane and the stereocilia of the OHCs (Allen, 1980; Neely and Kim, 1986). The transfer impedance $Z_{OHC}(x;\omega)$ then describes the space-frequency distribution of activity. The expression that we have used is of a somewhat simpler form than that used by Neely and Kim, namely,

$$Z_{OHC}(x;\omega) = c_0\, \omega_{loc}(x)\, \frac{1+i\beta(x;\omega)}{\delta_{SC} + i[\beta(x;\omega) - \sigma^2 / \beta(x;\omega)]}. \tag{2.9}$$

The resonance radian frequency of the BM, $\omega_{loc}(x)$, follows from equating the imaginary part of the passive BM impedance (Eq. 2.5) to zero; $\beta(x;\omega)$ is given by

$$\beta(x;\omega) = \omega / \omega_{loc}(x). \tag{2.10}$$

The numerical values of the new parameters are $c_0 = 0.06$ [kg m^{-2}], $\delta_{SC} = 0.14$ and $\sigma = 0.7$. The parameter σ indicates how much the OHC resonance has been shifted with respect to the BM resonance. For simplicity we have taken σ as constant so that the frequency-place map for the OHCs is parallel to that for the BM. The value we have chosen for σ is based on anatomical considerations (cf. Strelioff, 1986).

Solid lines in Fig. 1 illustrate BM resistance (thick lines in part a), BM reactance (thin lines in part a) and velocity response (part b) of a linear locally-active model. The BM impedance is given by Eq. (1.8); in this case positive as well as negative values occur in the resistance component. Where the resistance component is negative, energy is injected into the system - this happens in the region to the left of the response peak - and the BM response is amplified, see the solid line in part (b). The effect of the OHC activity

21

on the imaginary component of the BM impedance is mainly that of a basal shift of the place of resonance. We observe an enhancement of the velocity response of nearly 40 dB which is consistent with mechanical data found by Sellick *et al.* (1982) and Johnstone *et al.* (1986).

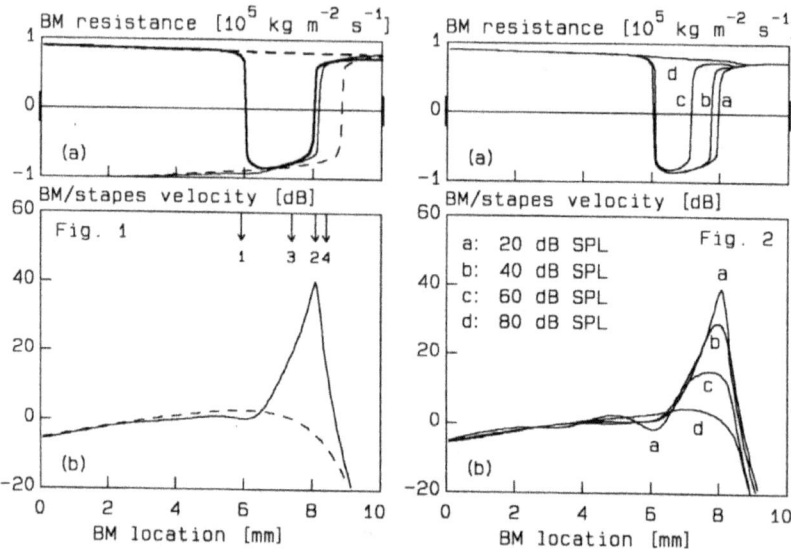

Figure 1. Model results for a linear passive and a locally-active long-wave model. Abscissa: Location x along the BM. The dashed lines correspond to the passive model with the BM impedance from Eq. (2.5). The solid lines belong to the locally-active model with the transfer function from Eq. (2.9). (a) The resistance (thick lines) and reactance (thin lines) components of the BM impedance drawn logarithmically for absolute values larger than 1×10^1 [kg m^{-2} s^{-1}]. The small bars on the left and the right indicate the region within which the impedance components are drawn in a linear way (with a slope of 0.043 to retain continuity). (b) BM velocity responses in dB normalized with respect to the stapes velocity (the arrows are explained in the legend to Fig. 3). Frequency 6 kHz, 500 sections.

Figure 2: Results of the quasi-linear locally-active model at four different input levels indicated in the figure. (a) The resistance component of the BM impedance. See also legend to Fig. 1. (b) BM velocity responses in dB normalized with respect to the stapes velocity. Frequency 6 kHz, 500 sections, 8 iterations.

At locations where the resistance component is negative the active structures (if considered in isolation) are zero-point unstable. Therefore, stability of the overall system is checked by considering the real part of the input impedance of the model; it has to be positive for all frequencies (if negative, energy would flow out of the model and our cochlear model would spontaneously emit sound). With the parameters given above, the system is stable. When c_0 is made larger than 0.06, the model becomes unstable.

2.3.4 The nonlinear locally-active case

We let the OHC pressure become a compressive function of movements of the BM. As a result the waveform of the BM response $v_{BM}(x,t)$ becomes distorted too, and the relation between pressure and velocity cannot be represented any more in terms of an impedance function. Because we want to solve our model in the frequency domain, we split all system variables into their Fourier components, and treat the distortion products as perturbations. In this section we consider only the primary components; in Appendix B we compute the third-order component in the BM velocity, and show that the coupling to the primary components is minimal.

We consider OHC transduction as consisting of two parts, a linear part according to Eq. (2.7) which describes the filtering of the BM velocity by the TM-cilia resonance, and a nonlinear part. The output of the linear part (which is the input to the nonlinear part), $I(x,t)$, is the time-varying counterpart of $P_{OHC}(x;\omega)$ from Eq. (2.7) divided by the scaling constant P_0 (which has the dimension of a pressure):

$$ I(x,t) = \left| Z_{OHC}(x;\omega) \, v_{BM}(x;\omega) \right| \sin(\omega t + \varphi_V(x;\omega) + \varphi_Z(x;\omega)) / P_0 . \tag{2.11} $$

Here $\varphi_V(x;\omega)$ and $\varphi_Z(x;\omega)$ denote the phases of $v_{BM}(x;\omega)$ and $Z_{OHC}(x;\omega)$, respectively. The complex variable $v_{BM}(x;\omega)$ (which will be called the *primary BM-velocity response)* is the first-order Fourier component of $v_{BM}(x,t)$, defined by

$$ v_{BM}(x;\omega) = -2i \int_0^T \frac{dt}{T} \, v_{BM}(x,t) \exp(i\omega t), \tag{2.12} $$

where T is equal to $2\pi/\omega$. A similar expression holds for the primary component of the pressure. The second stage is decribed by

$$ P_{OHC}^{NL}(x,t) = P_0 f[I(x,t)], \tag{2.13} $$

where $P^{NL}_{OHC}(x,t)$ is the nonlinear time-dependent OHC pressure, and where f[.] is a compressive function which is equal to the argument for very small argument values. The parameter P_0 scales the transition between linear and nonlinear behaviour of the time-varying pressure $P^{NL}_{OHC}(x,t)$.

The first Fourier component of the OHC pressure waveform $P_{OHC,NL}(x,t)$, denoted by $P_{OHC}(x;\omega)$, is defined by

$$ P_{OHC}(x;\omega) = -2i \int_0^T \frac{dt}{T} \, P_{OHC,NL}(x,t) \exp(i\omega t), \tag{2.14} $$

so that substitution of Eqs. (2.11) and (2.12) into Eq. (2.14) gives

$$P_{OHC}(x;\omega) = 2P_0 \exp(-i(\varphi_V(x;\omega) + \varphi_Z(x;\omega))) \times$$

$$\int_0^T \frac{dt}{T} \, f[|Z_{OHC}(x;\omega) \, v_{BM}(x;\omega) \, / \, P_0|\sin(\omega t)] \sin(\omega t). \qquad (2.15)$$

The quasi-linear impedance $Z^{QL}_{OHC}(x;\omega)$ defined by

$$Z^{QL}_{OHC}(x;\omega) \equiv P_{OHC}(x;\omega)/v_{BM}(x;\omega), \qquad (2.16)$$

is used to modify the passive BM impedance from $Z^p_{BM}(x;\omega)$ to $Z^{QL}_{BM}(x;\omega)$:

$$Z^{QL}_{BM}(x;\omega) = Z^p_{BM}(x;\omega) - Z^{QL}_{OHC}(x;\omega), \qquad (2.17)$$

The essence of the quasi-linear method is to use $Z^{QL}_{BM}(x;\omega)$ in a linear frequency-domain model. An iterative procedure must be used to solve the model because $Z^{QL}_{OHC}(x;\omega)$ depends on the BM velocity distribution which we do not know in advance. To start our computations we set $Z^{QL}_{OHC}(x;\omega)$ equal to $Z_{OHC}(x;\omega)$ at all locations x. This gives rise to an initial BM velocity distribution. With this distribution we adjust the quasi-linear impedance for the next iteration step, and so on.

Although the actual OHC transfer function is asymmetrical, we have taken for f[.] the hyperbolic tangent function. We chose this symmetrical function because we wanted to study amplitude compression. The scaling constant P_0 has been given the value 2 [N m^{-2}]. In our calculations the error in the velocity response between two subsequent iteration steps was negligible after eight iterations; in the integration we used twenty-four samples per period. In order to improve convergence, we averaged the velocities over two subsequent iteration steps. Appendix B shows that adding a third-order term to Eq. (2.11) does not noticeably influence the primary velocity response.

We developed our method on an Olivetti M28 computer (an antediluvian PC/AT with 8 MHz clock frequency) equipped with an 80287 coprocessor. Because the method is so fast we were able to calculate on this computer the BM response for a model with 500 sections in less than one minute. (With a 486 machine with clock frequency of 33 MHz it is a matter of seconds.)

2.4 Applications of the method

2.4.1 Nonlinear effect on the BM response pattern

Figure 2 presents results of the quasi-linear method applied to the nonlinear model for four input pressure levels, the layout being similar to that of Fig. 1. The BM resistance - part (a) of the figure - suffers clearly from de-activation as input level increases; eventually it approaches the passive resistance (cf. Fig. 1a). Note that de-activation is largest near the BM response peak. One effect of de-activation is self-suppression: The BM response - part (b) of the figure - flattens till it reaches the passive response (cf. Fig. 1b). Another effect is that the peak shifts to the left as the input level increases. Taking into account the relation between frequency and place domain, this is in agreement with the experimental results of Johnstone *et al.* (1986, their Fig. 4). For intermediate input levels the BM response becomes somewhat rippled due to a substantial retrograde wave originating from the region of the peak. Incomplete annihilation of individual retrograde wavelets is the reason that this retrograde wave is significant. Increasing the number of sections does not influence the result.

2.4.2 Input-output functions

The absolute value of the BM velocity at one location in the cochlea can be plotted as a function of input pressure. This produces an input-output (I/O) function. Four theoretical I/O curves obtained from the results of Fig. 2(b) are shown in Fig. 3; curve 1 pertains to the passive (and thus linear) model, the other curves to the locally-active (and thus nonlinear) model. The numbered arrows in Fig. 1(b) indicate the locations at which the I/O curves have been obtained. Arrow 1 points to the peak of the passive response and curve 1 is the I/O curve for the passive model; it is linear, of course. Curve 2 is the I/O function for the locally-active nonlinear model at the location of the peak; this curve is highly nonlinear and has a slope with a minimum value of about 0.3 [dB/dB], somewhat larger than the value of zero found by Johnstone *et al.* (1986, Fig. 5) and by Patuzzi *et al.* (1984a, Fig. 1). Curve 3 is obtained for locations to the left of the peak; in this case the nonlinearity of the BM response is less pronounced than for curve 2. Besides, curve 3 intersects with curve 2. These results are consistent with results of mechanical experiments (e.g., Sellick *et al.*, 1982, Fig. 5; Robles *et al.*, 1986a, Figs 1 and 2) and with those of intracellular recordings in inner hair cells (Russell and Sellick, 1978, Fig. 3). For locations to the right of the peak (curve 4) the I/O curve is less nonlinear than at the peak; this does not agree with the mechanical data.

2.5 Discussion

We have described a quasi-linear method to solve nonlinear cochlear models. The method can be applied to any stable locally-active nonlinear model if the system is linear at low levels and saturates at high levels. We use our method to compute responses of a long-wave model in which activity is place-frequency dependent and nonlinearity is associated solely with activity. We believe that the quasi-linear method is more useful than a time-domain approach for the following reasons: (a) The familiar concept of impedance is extended to become a natural element of the quasi-linear method. The impedance can be examined at every location in the cochlea and consequences of its variations can be determined or judged easily. (b) *Gedanken* experiments that do not have a counterpart in the time domain are possible. For instance, impedances can be manipulated to such an extent that the Hilbert relation between real and imaginary parts is violated so that they become physically unrealizable (yet the model remains stable). (c)

In the quasi-linear method every variable can be interpreted directly but to asses waveform distortion in the time-domain method is difficult. (d) The quasi-linear method is much faster than existing solution methods in the time domain. This feature makes it easier to acquire insight into the behaviour of different locally-active models for various parameter values. However, the time-domain method has undeniable advantages when using multi-component, impulsive or non-stationary signals as stimuli.

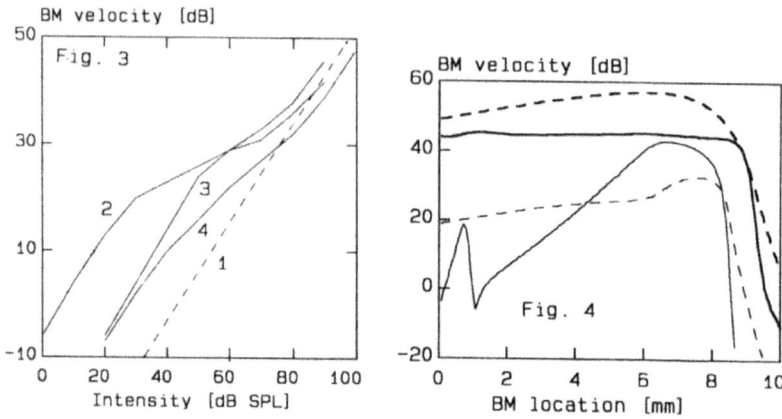

Figure 3: Four theoretical I/O functions for pure tone stimulation. The dashed curve (curve 1) has been obtained at the response peak in the passive linear model (see dashed response pattern and arrow 1 in Fig. 1b). The other curves (curves 2-4) belong to the locally-active nonlinear model, and have been obtained at the corresponding arrows in Fig. 1(b). The reference value (0 dB) of the ordinate corresponds to a velocity of 1×10^{-5} [m s^{-1}]. Frequency 6 kHz, 500 sections, 8 iterations.

Figure 4: Primary (dashed lines) and third-order (solid lines) velocity response for two primary input levels (thin lines: 70 dB SPL; thick lines: 100 dB SPL). The reference value of the primary BM velocity is 1×10^{-5} [m s^{-1}], for the third-order BM velocity it is 1×10^{-7} [m s^{-1}] (40 dB difference). The thin lines correspond to the 'worst-case' scenario in which a minimal difference of only 25 dB between the primary and third-order velocity response is found. Primary frequency 6 kHz, 500 sections, 8 iterations for the primary response, 3 iterations for third-order response.

We will illustrate point (b) with an example. The theoretical I/O functions shown in this paper have a minimal slope in the nonlinear region of about 0.3 [dB/dB]. This value is larger than the value of zero found by Johnstone et al. (1986, Fig. 5) and by Patuzzi et al. (1984a, Fig. 1). To achieve a zero slope in the model we need to produce a response peak that shifts more to the left with increasing stimulus intensity than does the peak in Fig. 1(b). This can indeed be obtained with an OHC transfer impedance that differs from that of Eq. (2.9). However, it turns out that the new transfer impedance is not physically realizable. As far as we know, a similar shift does not occur in an existing realizable locally-active model. This *Gedanken* experiment would not have been possible with the time-domain method.

The influence of decreasing activity on the BM response has been examined by other authors (Neely and Kim, 1986; Neely, 1993; Geisler, 1991). In that work activity was made to decrease by a global factor independent of cochlear location. Our analysis is more realistic (as far as nonlinearity is concerned) in that de-activation depends on the (place-dependent) BM velocity distribution.

Mountain and Hubbard (1983), Patuzzi *et al.* (1989) and Yates (1990) studied a saturating nonlinearity that did not form a part of a cochlear model. Consequently, they could only assess the local and not the global effect[3] of the nonlinearity on the BM response. Neither could any one of these authors predict possible consequences on wave reflection in the cochlea.

In the models by Hall (1974, 1977) and Strube (1986) the nonlinearity affected only the resistance component of the BM impedance. In Hall's work damping increases with increasing stimulation, in Strube's case it is undamping which decreases in magnitude. In both cases the effect is independent of frequency. As can be derived from the elaborations of de Boer (1983) this is not the right way to achieve gradual cochlear amplification. In more modern and realistic models local activity is thought to be due to a bi-directional coupling mechanism (Kim et al. 1980b; Weiss, 1982) which includes a certain amount of filtering. In that case nonlinearity will modify the magnitude of the transfer impedance involved and thus affect the imaginary as well as the real part of that impedance. This is also true in the model we describe in this paper. Note that with our technique a *Gedanken* experiment is possible in which the roles played by real and imaginary parts of this impedance are separated.

The quasi-linear method can easily be extended to phenomena like suppression and production of combination tones. The OHC pressure will then be distorted by a secondary tone of a different frequency. These extensions will be treated in future papers; for an application to combination tones see Kanis and de Boer (1993). In fact, the original reason for undertaking this work was to study the relation between two-tone suppression and generation of combination tones.

Acknowledgement

This work was supported by the Netherlands Organization for Pure Research. The authors thank C.R. Steele and an anonymous reviewer for their valuable comments.

---

[3] That in linear models local and global effects can be separated is apparent from the LG approximation (Zweig *et al.*, 1976). The LG expression consists of two factors, a local factor depending only on the BM impedance and a global one containing an integral of the local wavenumber $k(x)$ over location x. The former factor approximately describes the shape of the BM velocity response to the left of the peak. The latter factor describes where the propagating wave changes into an evanescent wave (cf. de Boer, 1980). We can consider an LG approximation in every step of our iteration procedure determined by the quasi-linear BM impedance $Z_{BM,QL}(x;\omega)$ of Eq. (17). By analogy we can recognize the same local and global effects.

Appendix 2.A: Coupling to the middle ear

We couple the stapes to the outside world with an impedance Z_0 to ensure that the model remains stable for all frequencies. The local impedance in the cochlea can be solved from a Riccati equation (Kaernbach et al., 1987). A good approximation to this impedance is derived from the Liouville-Green (LG) solution (also known as WKB solution) of the wave equation (Viergever and de Boer, 1987); it is the conjugate of this impedance which we use for Z_0 (de Boer, 1991). In this way we have optimal transfer of acoustical power - in fact, we have modelled an ideal middle ear. The equation describing the coupling is

$$g\, p_0 - p(0) = -Z_0 v_{stapes},$$
(2.A.1)

where g is a scaling factor, p_0 the given input pressure and v_{stapes} the stapes velocity which relates to the derivative $p_x(x)$ of the pressure as

$$v_{stapes} = -p_x(0) / (i\omega\rho).$$
(2.A.2)

For the purpose of the present paper where only one frequency (6 kHz) is considered it is not necessary to model the middle ear in any further detail. The coefficient g has been given a value of 3.5×10^2; then, for an input pressure of 14 dB SPL the velocity response of our locally-active model at its peak has a magnitude of 4×10^{-5} m s^{-1} which is equal to the value obtained by Sellick et al. (1982, Fig. 15, curve with closed circles).

Appendix 2.B: Justification of the method

The quasi-linear approach is valid if the harmonic distortion does not influence the primary response. To check this, we calculate the third-order velocity response (this distortion product being the most important when using a hyperbolic tangent function), and evaluate its effect on the primary response. The third-order response is denoted by $v_{BM}(x;3\omega)$ and given by

$$v_{BM}(x;3\omega) = -2i \int_0^T \frac{dt}{T} v_{BM}(x,t) \exp(3i\omega t),$$
(2.B.1)

which is analogous to Eq. (2.12) of the main text. Similarly, we define the third-order pressure response $p(x;3\omega)$ by

$$p(x;3\omega) = -2i \int_0^T \frac{dt}{T} p(x,t) \exp(3i\omega t).$$
(2.B.2)

On the assumption that only long waves exist in the cochlea, the relation between BM velocity and pressure is given by

$$p_{xx}(x;\omega) = -i\omega\rho\, v_{BM}(x;\omega)\,/\,h. \tag{2.B.3}$$

This equation is true for any frequency; for the primary frequency $\omega/2\pi$ it leads to the wave equation (Eq. 2.1) and for the third harmonic to an inhomogeneous equation as we shall see.

Third-order distortion is generated whenever the BM vibrates with such a large amplitude that the OHC pressure becomes nonlinear. At every point in the cochlea we define, in analogy to Eq. (2.B.2), the third-order component $P_{OHC}(x;3\omega)$ of the OHC pressure $P^{NL}_{OHC}(x,t)$ [defined in Eq. (2.13)] as

$$P_{OHC}(x;3\omega) = -2i \int_{0}^{T} \frac{dt}{T}\, P_{OHC,NL}(x,t)\, \exp(3i\omega t). \tag{2.B.4}$$

Because the OHCs also respond to oscillations at frequency $3\omega/2\pi$, the argument $I(x,t)$ of the nonlinear function f[.] contains two terms:

$$I(x,t) = \left|Z_{OHC}(x;\omega)\, v_{BM}(x;\omega)\,/\,P_0\right|\, \sin(\omega t + \varphi_V(x;\omega) + \varphi_Z(x;\omega)) +$$
$$\left|Z_{OHC}(x;3\omega)\, v_{BM}(x;3\omega)\,/\,P_0\right|\, \sin(3\omega t + \varphi_V(x;3\omega) + \varphi_Z(x;3\omega)). \tag{2.B.5}$$

The phases $\varphi_V(x;3\omega)$ and $\varphi_Z(x;3\omega)$ are the phases of the third-order response and $Z_{OHC}(x;3\omega)$, respectively.

For the third-order distortion product the equilibrium over the BM is described by

$$Z^{p}_{BM}(x;3\omega)\, v_{BM}(x;3\omega) = -2p(x;3\omega) + P_{OHC}(x;3\omega). \tag{2.B.6}$$

Eliminating $v_{BM}(x;3\omega)$ from Eq. (2.B.6) with Eq. (2.B.3) evaluated at frequency $3\omega/2\pi$, we obtain an inhomogeneous wave equation

$$p_{xx}(x;3\omega) + k^2(x;3\omega)\, p(x;3\omega) = 0.5\, k^2(x;3\omega)\, P_{OHC}(x;3\omega) \tag{2.B.7}$$

The factor $k^2(x;3\omega)$ is defined as in Eq. (2.2) but evaluated at frequency $3\omega/2\pi$. The third-order pressure response is solved from Eq. (2.B.7), and $v_{BM}(x;3\omega)$ is calculated from Eq. (2.B.6).

It is important to realize that the solution of the inhomogeneous wave equation proceeds with a modified boundary condition at $x = 0$, because for the third harmonic p_0 in Eq. (2.A.1) is zero. In order to evaluate Eq. (2.B.5) we need to know the primary as well as the third-order velocity response: First the primary response is computed in eight iteration steps, and then the third-order response is computed in three iteration steps, using the modified boundary condition as described above.

Figure 4 shows primary (dashed lines) and third-order (solid lines) velocity responses for two primary input levels of 70 and 100 dB SPL. The third-order response has been amplified by 40 dB with respect to the primary response. For an input level of 70 dB SPL the largest amount of nonlinearity occurs in the region of the peak of the primary response, while for an input level of 100 dB SPL third-order distortion is also found at places to the left and the right of the primary peak. The maximal amount of distortion corresponds to the minimal difference between the primary and the third-order velocity response. For a primary input level of 70 dB SPL this minimal difference is about 25 dB. Thus, third-order distortion is always *more* than 25 dB below the primary velocity response. However, one factor has not yet been taken into account, namely, the possible influence of the third-order response on the primary response. To assess this influence, we modify Eq. (2.11) to accomodate the input given by Eq. (2.B.5), and use Eqs. (2.14), (2.16) and (2.17) to solve the wave equation for the primary BM response. The resulting BM response turns out to be indistinguishable from the first-order responses shown in Fig. 2 (the difference is below 0.01 dB). This procedure is repeated for fifth and higher-order distortion products which yields the same result. Thus, higher-order products do not influence the primary response. Note that in our model the perturbation method is valid for all intensities, since for large input levels the nonlinear term in our system equations vanishes in a relative sense.

References

Allen, J.B. (1977). "Two-dimensional cochlear fluid model: New Results," J. Acoust. Soc. Am. 61, 110-119.

Allen. J.B. (1980). "Cochlear micromechanics - A physical model of transduction," J. Acoust. Soc. Am. 68, 1660-1670.

Allen. J.B. (1990). "Modeling the noise damaged cochlea," in: *The Mechanics and Biophysics of Hearing*, edited by P. Dallos, C.D. Geisler, J.W. Matthews, M.A. Ruggero and C.R. Steele (Springer, Berlin), pp. 324-331.

Ashmore, J.F. (1987). "A fast motile response in guinea-pig outer hair cells: The cellular basis for the cochlear amplifier," J. Physiol. 388, 323-347.

Boer, E. de (1980). "Auditory Physics. Physical principles in hearing theory. I," Phys. Rep. 62, 87-174.

Boer, E. de (1982). "A correspondence principle in cochlear mechanics," J. Acoust. Soc. Am. 71, 1496-1501.

Boer, E. de (1983). "No sharpening? A challenge for cochlear mechanics," J. Acoust. Soc. Am. 73, 567-573.

Boer, E. de (1984). "Auditory Physics. Physical principles in hearing theory. II," Phys. Rep. 105, 141-226.

Boer, E. de(1991). "Auditory Physics. Physical principles in hearing theory. III," Phys. Rep. 203, 125-231.

Cody, A.R., and Russell, I.J. (1987). "The responses of hair cells in the basal turn of the guinea pig cochlea to tones," J. Physiol. 383, 551-569.

Duifhuis, H., Hoogstraten, H.W., van Netten, S.M., Diependaal, R.J. and Bialek, W. (1985). "Modelling the cochlear partition with coupled Van der Pol oscillators," in: *Peripheral Auditory Mechanisms*, edited by J.B. Allen, J.L. Hall, A. Hubbard, S.T. Neely and A. Tubis (Springer, Berlin), pp. 290-297.

Geisler, C.D. (1991). "A cochlear model using feedback from motile outer hair cells," Hear. Res. 54, 105-117.

Gitter, A.H. and Zenner, H.-P. (1988). "Auditory transduction steps in single inner and outer hair cells," in: *Basic issues in Hearing*, edited by H. Duifhuis, J.W. Horst and H.P. Wit (Academic Press, London), pp. 32-39.

Hall, J.L. (1974). "Two-tone distortion products in a nonlinear model of the basilar membrane," J. Acoust. Soc. Am. 56, 1818-1828.

Hall, J.L. (1977). "Two-tone suppression in a nonlinear model of the basilar membrane," J. Acoust. Soc. Am. 61, 802-810.

Hubbard, A.E., and Geisler, C.D. (1972). "A hybrid-computer model of the cochlear partition," J. Acoust. Soc. Am. 51, 1895-1903.

Hudspeth, A.J. and Corey, D.P. (1977). "Sensitivity, polarity, and conductance change in the response of vertebrate hair cells to controlled mechanical stimuli," Proc. Natl. Acad. Sci. USA 74, 2407-2411.

Johnstone, B.M., Patuzzi, R. and Yates, G.K. (1986). "Basilar membrane measurements and the travelling wave," Hear. Res. 22, 147-153.

Kaernbach, Chr., König, P. and Schillen, Th. (1987). "On Riccati equations describing impedance relations for forward and backward excitation in the one-dimensional cochlea model," J. Acoust. Soc. Am. 81, 408-411.

Kanis, L.J., and Boer, E. de (1993). "The emperor's new clothes: DP emissions in a locally-active nonlinear model of the cochlea," in: *Biophysics of Hair Cell Sensory Systems*, edited by H. Duifhuis, J.W. Horst, P. van Dijk, and S.M. van Netten (World Scientific, Singapore), pp. 304-311.

Kemp, D.T. (1979). "Evidence of mechanical nonlinearity and frequency selective wave amplification in the cochlea," Arch. Otorhinolaryngol. 224, 37-45.

Khanna, S.M. and Leonard, D.G.B. (1982). "Basilar membrane tuning in the cat cochlea," Science 215, 305-306.

Kim, D.O., Molnar, C.E. and Pfeiffer, R.R. (1973). "A system of nonlinear differential equations modeling basilar-membrane motion," J. Acoust. Soc. Am. 54, 1517-1529.

Kim, D.O., Neely, S.T. , Molnar, C.E. and Matthews, J.W. (1980a). "An active cochlear model with negative damping in the partition: Comparison with Rhode's ante- and post-mortem observations," in: *Psychophysical, Physiological and Behavioural Studies in Hearing*, edited by G. van den Brink and F.A. Bilsen (Delft Univ. Press, Delft), pp. 7-14.

Kim, D.O., Molnar, C.E. and Matthews, J.W. (1980b). "Cochlear mechanics: Nonlinear behavior in two-tone responses as reflected in cochlear-nerve-fiber responses and in ear-canal sound pressure," J. Acoust. Soc. Am. 67, 1704-1721.

Kim, D.O. (1986). "A review of nonlinear and active cochlear models", in: *Peripheral Auditory Mechanisms*, edited by J.B. Allen, J.L. Hall, A. Hubbard, S.T. Neely and A. Tubis (Springer, Berlin), pp. 239-249.

Le Page, E.L. and Johnstone, B.M. (1980). "Nonlinear mechanical behaviour of the basilar membrane in the basal turn of the guinea pig cochlea," Hear. Res. 2, 183-189.

Lighthill, M.J. (1981). "Energy flow in the cochlea," J. Fluid. Mech. 106, 149-213.

Matthews, J.W. (1980). *Mechanical modeling of nonlinear phenomena observed in the peripheral auditory system* (Washington University, St. Louis, Missouri).

Mountain, D.C., Hubbard, A.E. and McMullen, T.A. (1983). "Electromechanical processes in the cochlea," in: *Mechanics of Hearing*, edited by E. de Boer and M.A. Viergever (University Press, Delft), pp. 119-126.

Neely, S.T. and Kim, D.O. (1983). "An active cochlear model showing sharp tuning and high sensitivity," Hear. Res. 9, 123-130.

Neely, S.T. and Kim, D.O. (1986). "A model for active elements in cochlear biomechanics," J. Acoust. Soc. Am. 79, 1472-1480.

Neely, S.T. (1993). "A model of cochlear mechanics with outer hair cell motility," submitted to J. Acoust. Soc. Am.

Nuttal, A.L., Dolan, D.F. and Avinash, G. (1990). "Measurements of basilar membrane tuning and distortion with laser doppler velocimetry," in: *The Mechanics and Biophysics of Hearing*, edited by P. Dallos, C.D. Geisler, J.W. Matthews, M.A. Ruggero and C.R. Steele (Springer, Berlin), pp. 288-295.

Patuzzi, R.B., Johnstone, B.M. and Sellick, P.M. (1984a). "The alteration of the vibration of the basilar membrane produced by loud sound," Hear. Res. 13, 99-100.

Patuzzi, R.B., Sellick, P.M. and Johnstone, B.M. (1984b). "The modulation of the sensitivity of the mammalian cochlea by low frequency tones. III. Basilar membrane motion," Hear. Res. 13, 19-27.

Patuzzi, R.B., Yates, G. K. and Johnstone, B.M. (1989). "Outer hair cell receptor current and sensorineural hearing loss," Hear. Res. 42, 47-72.

Rhode, W.S. (1971). "Observations of the vibration of the basilar membrane in squirrel monkeys using the Mössbauer technique," J. Acoust. Soc. Am. 49, 1218-1231.

Rhode, W.S. (1977). "Some observations on two-tone interaction measured with the Mössbauer effect," in: *Psychophysics and Physiology of Hearing*," edited by E.F. Evans and J.P. Wilson (Academic Press, London), pp. 27-38.

Rhode, W.S. (1978). "Some observations on cochlear mechanics", J. Acoust. Soc. Am. 64, 158-176.

Robles, L., Ruggero, M.A. and Rich, N.C. (1986a). "Basilar membrane mechanics at the base of the chinchilla cochlea. I. Input-output functions, tuning curves, and response phases," J. Acoust. Soc. Am. 80, 1364-1374.

Robles, L., Ruggero, M.A. and Rich, N.C. (1986b). "Mössbauer measurements of the mechanical response to single-tone and two-tone stimuli at the base of the chinchilla cochlea," in: *Peripheral Auditory Mechanisms*, edited by J.B. Allen, J.L. Hall, A. Hubbard, S.T. Neely and A. Tubis (Springer, Berlin), pp. 121-128.

Robles, L., Ruggero, M.A. and Rich, N.C. (1990). "Two-tone distortion products in the basilar membrane of the chinchilla cochlea," in: *The Mechanics and Biophysics of Hearing*, edited by P. Dallos, C.D. Geisler, J.W. Matthews, M.A. Ruggero and C.R. Steele (Springer, Berlin), pp. 304-311.
Robles, L., Ruggero, M.A. and Rich, N.C. (1991). "Two-tone distortion in the basilar membrane of the cochlea," Nature 349, 413-414.

Ruggero, M.A., and Rich, N.C. (1991). "Application of a commercially-manufactured Doppler-shift laser velocimeter to the measurement of basilar-membrane vibration," Hear. Res. 51, 215-230.

Ruggero, M.A., Robles, L. and Rich, N.C. (1992). "Two-tone suppression in the basilar membrane of the cochlea: Mechanical basis of auditory-nerve rate suppression," J. Neurophysiol. 68, 1087-1099.

Russell, I.J. and Sellick, P.M. (1978). "Intracellular studies of hair cells in the mammalian cochlea," J. Physiol. 284, 261-290.

Sellick, P.M., Patuzzi, R. and Johnstone, B.M. (1982). "Measurement of basilar membrane motion in the guinea pig using the Mössbauer technique," J. Acoust. Soc. Am. 72, 131-141.

Strelioff, D. (1986). "Role of passive mechanical properties of outer hair cells in determination of cochlear mechanics", in: *Peripheral Auditory Mechanisms*, edited by J.B. Allen, J.L. Hall, A. Hubbard, S.T. Neely and A. Tubis (Springer, Berlin), pp. 239-249.

Strube, H.W. (1985). "A computationally efficient basilar-membrane model," Acustica 58, 207-214.

Strube, H.W. (1986). "The shape of the nonlinearity generating the combination tone $2f_1$-f_2," J. Acoust. Soc. Am. 79, 1511-1518.

Viergever, M.A. and de Boer, E. (1987). "Matching impedance of a nonuniform transmission line: Application to cochlear modeling," J. Acoust. Soc. Am. 81, 184-186.

Weiss, T.F. (1982). "Bidirectional transduction in vertebrate hair cells: A mechanism for coupling mechanical and electrical processes," Hear. Res. 7, 353-360.

Yates, G.K. (1990). "The basilar membrane nonlinear input-output function," in: *The Mechanics and Biophysics of Hearing*, edited by P. Dallos, C.D. Geisler, J.W. Matthews, M.A. Ruggero and C.R. Steele (Springer, Berlin), pp. 106-113.

Zurek, P.M. (1981). "Spontaneous narrowband acoustic signals emitted by human ears," J. Acoust. Soc. Am. 69, 514-523.

Zweig, G., Lipes, R. and Pierce, J.R. (1976). "The cochlear compromise," J. Acoust. Soc. Am. 59, 975-982.

Zwicker, E. (1986). "A hardware cochlear nonlinear preprocessing model with active feedback," J. Acoust. Soc. Am. 80, 146-153.

3

Two-tone suppression in a locally-active nonlinear model of the cochlea*)

Abstract   In auditory nerve, inner-hair cell and basilar-membrane responses, it has been found that the response to one tone can be suppressed by another tone. This phenomenon, called two-tone suppression, is examined on the level of the basilar membrane with a locally active long-wave model of the cochlea in which the active mechanism is nonlinear. The model is solved in the frequency domain by means of a quasilinear solution method. Several phenomena, such as difference in growth of suppression for low-side and high-side suppressors and critical dependence of the phase of the probe response on suppressor parameters, have been replicated. The attenuation hypothesis which states that the presence of a suppressor has the same effect on the probe response as attenuation of probe level is shown to be insufficient in explaining the experimental data. Our model, in which suppression is simply reduction of power amplification due to saturation of the active mechanism, is more successful in this respect.

3.1 Introduction

Since its discovery in cochlear microphonic recordings (Covell and Black, 1936), two-tone suppression has been studied extensively in responses of auditory-nerve fibers (Sachs and Kiang, 1968; Sachs, 1969; Arthur *et al.*, 1971; Abbas and Sachs, 1976; Sachs and Abbas, 1976; Javel *et al.*, 1978), inner hair cells (Sellick and Russell, 1979; Cheatham and Dallos, 1989, 1990; Nuttall and Dolan, 1993a), and the basilar membrane (BM) (Rhode, 1977; Patuzzi *et al.*, 1984; Robles *et al.*, 1986b; Ruggero *et al.*, 1992; Nuttall and Dolan, 1993a; Rhode and Cooper, 1993). During two-tone suppression the response to a probe tone is suppressed by a secondary tone that is present in the stimulus to the ear but not necessarily detectable in the measured response. Suppression of a probe tone has been observed for low-side (LS) suppressors as well as for high-side (HS)

---

*) Preliminary results were presented during the 3rd International Symposium on *Cochlear Mechanisms and Otoacoustic Emissions* in Rome, 1992.

suppressors where low-side and high-side stand for frequencies below and above the probe frequency, respectively.

Not only the magnitude but also the phase of the probe response changes during two-tone suppression. Several authors (Nuttall and Dolan, 1993a, 1993b; Cheatham and Dallos, 1989, 1990, 1993) have tried to explain the experimental phase data with the *attenuation hypothesis*. Sachs and Abbas (1976) introduced the concept of attenuation in a phenomenological model for neural rate functions but Javel *et al.* (1978) were the first to apply this idea also to phase measurements and to give the hypothesis its current name. The hypothesis states that addition of a suppressor to the stimulus has the same effect on the probe response as attenuation of probe level, with an amount that is only dependent on suppressor level.[1] Although it originally applied to probe tones with frequency equal to the characteristic frequency (CF), the hypothesis was extended to off-CF tones. In some cases the hypothesis correctly predicts response behavior, in other cases it does not (Geisler and Sinex, 1980; Deng and Geisler, 1985; Costalupes *et al.*, 1987; Nuttall and Dolan, 1993a, 1993b; Cheatham and Dallos, 1989, 1990, 1993).

We believe that any agreement of the experimental data with the attenuation hypothesis is only phenomenological and has no physical basis. In this paper we will try to convince the reader that two-tone suppression is described more realistically by *saturation* of the active mechanism. The idea is that a suppressor tone interferes with the (nonlinear) active process that enhances the probe response. This may happen for HS as well as for LS suppressors. For low-level HS suppressors the region of interference will be mostly basally to the peak, while low-level LS suppressors will mainly saturate the active process at locations at and apical to the probe peak. The idea of relating two-tone suppression to saturation of the active process is not new. Some authors have described the relation qualitatively (Kim, 1986; Patuzzi *et al.*, 1989; Geisler *et al.*, 1990; Cheatham and Dallos, 1990; Zweig, 1991); others have incorporated saturation of activity in a cochlea model (Zwicker, 1979, 1986; Mountain and Hubbard, 1983; Neely and Stover, 1993; Geisler *et al.*, 1993; Cohen and Furst, 1993) but two-tone suppression has never systematically been described with a nonlinear locally-active cochlea model.

In this paper we have examined the mechanism of two-tone suppression with a one-dimensional cochlea model in which frequency-specific activity and compressive nonlinearity both reside in the generation of pressures by outer hair cells (OHCs). To illustrate the relation between saturation of the active mechanism and response in an elegant way we have solved the model in the frequency domain with a *quasilinear* method which treats distortion products in the BM-velocity response as perturbations. For the purpose of this paper the model need only be solved for the primary components in the traveling wave; the computation of harmonics or combination tones (Kanis and de Boer, 1994) is optional because their influence on the primary components can be neglected (cf. Kanis and de Boer, 1993b).

---

[1] In a model by Sachs (1969) suppression is a function of the difference between probe and suppressor level, but this idea was rejected in a later study by Abbas and Sachs (1976).

3.2 Method

We have modeled the cochlea as a rectangular box divided into two fluid-filled scalae by a movable partition consisting of the basilar membrane (BM) and the organ of Corti. The model is driven at the stapes which is coupled to the pressure source by means of a reflectionless middle ear. Pure-tone stimulation results in a travelling wave on the BM that peaks at a location that depends on the frequency of the stimulus. For simplicity, we approximate the cochlear waves by long waves. Since passive models do not have realistic tuning properties (cf. de Boer, 1991) we have included active outer hair cells (OHCs) which locally generate sound pressures that are added to the pressure difference over the BM. At locations basal to the resonance place the OHCs inject more energy into a cochlear section than is absorbed by that section; at these locations the cochlea is *locally active*. The OHC-generated pressure causes amplification of the traveling wave which ultimately results in a 40 dB increase of the velocity response at the peak. We made the pressure generation by the OHCs a saturating function of the input to the OHCs which is in line with physiological data (cf. Hudspeth and Corey, 1977). Note that this is the only nonlinearity introduced into the model.

The response of the model is solved in the frequency domain by considering only the primary Fourier components in the pressure generated by the OHCs and neglecting the influence of other components. Because the Fourier components depend nonlinearly on the BM response the final solution is obtained after a number of iteration steps. In each iteration step a linear problem is solved so that the concept of BM impedance can be used throughout. In our case this impedance includes effects of the active mechanism (and its saturation), and for this reason we will speak of it as an *effective* BM impedance. For details about the solution method, its validity, and model parameters, the reader is referred to Appendix A of this paper, and to Kanis and de Boer (1993b).

3.3 Results

3.3.1 Suppression and self-suppression: BM-velocity responses

In Fig. 1 the effect of saturation of the active mechanism on the BM response is illustrated for *single-tone* stimulation at 7 kHz. Model results are shown as a function of location x (distance from the stapes) for three different input levels. Part (a) shows the resistance component of the effective BM impedance (the BM impedance that describes both the passive and active properties of the cochlear partition). We refer to the legend for details about the scaling. To avoid cluttering the figure the reactance component of the BM impedance has been omitted. Part (b) shows the magnitude of the BM-velocity response normalized to stapes velocity, and part (c) shows the phase of the BM-velocity response. Input levels (at the eardrum) are 30, 65 and 100 dB SPL for the thick solid, thin solid and thick dashed lines, respectively. At 30 dB SPL a pronounced excursion of the BM-resistance component into the negative domain is seen. It is located basally to the response peak. In this case the model is almost linear. At higher levels of stimulation nonlinearity comes into play because the active elements are getting saturated: the negative lobe of the BM resistance becomes less prominent and, consequently, the BM-velocity response in the peak region is reduced. At 100 dB SPL the negative-going excursion has disappeared completely, and the BM response approximates that of a passive model. The phenomenon that in the peak region the BM response suppresses itself by bringing the active mechanism into saturation is called *self-suppression*.

It should be noted that the effective BM impedance gives a *relative* measure of the generated pressures: the OHC-generated pressures increase for increasing input levels (see also Cohen and Furst, 1993, Fig. 2) in spite of the fact that the negative lobe in Fig. 1(a) shrinks. Another comment concerns the fact that the largest pressures are generated at the *peak* of the response since the pressures are the product of the BM-impedance and the velocity response [see Eq. (1.A.10) for the linear case]. Therefore, one should not say that the region of amplification lies basally to the peak. It is only if one looks at the energy flow, that at regions basal to the peak, where the resistance is negative, more energy is injected into the cochlea than absorbed.

Phase behavior in Fig. 1(c) at increasing input levels is that of a lag (lead) at locations basal (apical) to the peak and no phase change at the peak location of 30 dB SPL (in fact, the transition point lies 0.1 mm basally to the peak). Bearing in mind the correspondence between locations basal (apical) to the peak and frequencies below (above) the characteristic frequency (CF), we find that the phase behavior in Fig. 1(c) is in accordance with experimental data (Anderson *et al.*, 1971; Rhode and Robles, 1974; Sellick *et al.*, 1982, Fig. 16; Dallos, 1986, Fig. 7; Cheatham and Dallos, 1989, Fig. 4, 1990, Fig. 11; Nuttall and Dolan, 1993a, Figs. 1 and 2).

In Fig. 1 we have illustrated how increasing the input level results in decreased amplification of the travelling wave. A similar decrease in amplification may occur when a secondary tone is added to the stimulus. Three examples of two-tone suppression will be shown, all for a 7 kHz probe presented at 30 dB SPL. We start with a suppressor frequency of 8 kHz in Fig. 2, then decrease it to 7.25 kHz in Fig. 4 and our last example in Fig. 5 has been obtained for a suppressor frequency of 5.5 kHz. We will see that the nature of suppression changes dramatically from one example to the other, while the mechanism of saturation remains the same.

In Fig. 2 an example of high-side (HS) suppression is illustrated in a similar layout as Fig. 1. The resistance function depicted in part (a) and the phase response in part (c) refer to the probe frequency. At the location of the unsuppressed probe-response peak (thin solid lines) the 8 kHz suppressor at 60 dB SPL (thick dashed line in part b) is smaller in amplitude than the suppressed probe (thin dashed lines). We will call this case *non-dominant* suppression. It is seen in part (a) that in the presence of the suppressor the negative lobe shrinks which results in a magnitude decrease (i.e., suppression) of the probe response at and around the probe peak. Furthermore, it is seen that this magnitude decrease results in a modest *release* of saturation in the region from 6.9 to 7.2 mm: the rightmost crossing of the dashed line in part (a) lies apically to the zero-crossing of the solid line.

In part (c) we observe phase lags at and apically to the peak (this is more clearly seen in Fig. 3 where phase shifts have been enlarged). These phase lags are consistent with the small but significant phase lags that have been observed in many studies of neural, inner hair cell and BM responses (Arthur *et al.*, 1971; Javel *et al.*, 1978; Deng and Geisler, 1985; Robles *et al.*, 1989; Cheatham and Dallos, 1989, 1990; Ruggero *et al.*, 1992; Nuttall and Dolan, 1993a). The phase lags at locations basal to the peak are consistent with the lags found by Cheatham and Dallos (1989, Figs. 7 and 11) and Nuttall and Dolan (1993b, Fig. 2b).

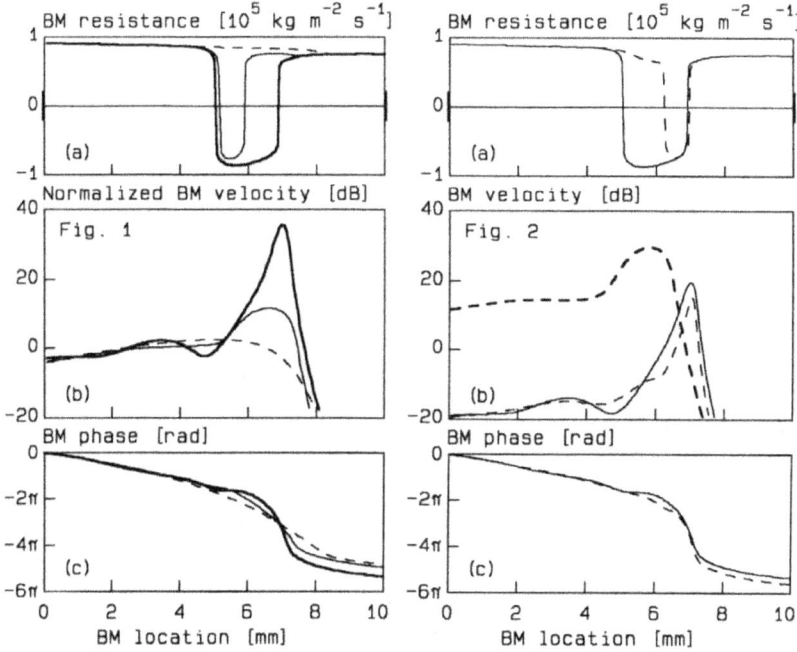

Figure 1. Single-tone results. The input stimulus consists of one 7 kHz tone presented at three different levels. Abscissa: Location x along the BM, measured from the stapes. (a) Resistance component of the BM impedance drawn logarithmically for absolute values larger than 10 [kg m^{-2} s^{-1}]. The small bars on the left and right indicate the region within which the resistance has been plotted in a linear way (with a slope of 0.043 to retain continuity). (b) Magnitude of BM-velocity responses normalized to stapes velocity. (c) Phase of BM-velocity responses. Input levels are 30, 65 and 100 dB SPL for the thick solid, thin solid and thick dashed lines, respectively. The figure illustrates how the amount of activity decreases (in a relative sense) when input level increases resulting in a flattening of the peak.

Figure 2. Non-dominant high-side suppression. (Non-dominant: at the location of the unsuppressed probe peak the suppressor has a smaller amplitude than the suppressed probe). A 7 kHz probe at 30 dB SPL is suppressed by a 8 kHz suppressor at 60 dB SPL. Solid lines pertain to single-tone, and dashed lines to two-tone stimulation. Thin dashed lines belong to the probe and thick dashed lines to the suppressor. (a) BM resistance at the probe frequency. (b) Velocity responses in dB relative to 1×10^{-5} m s^{-1}. (c) Phase of probe velocity response. Note in the two-tone case the increase of the phase lag at the location of the peak and at more apical locations.

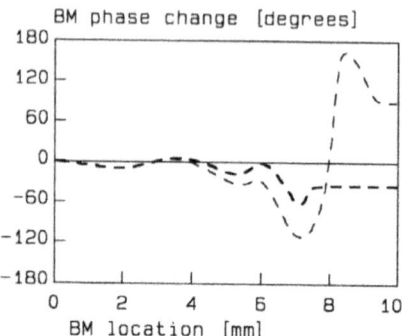

BM phase change [degrees]

BM location [mm]

Figure 3. BM-phase change with respect to the phase of the unsuppressed probe response at 7 kHz and 30 dB SPL (thick solid line in Fig. 1c). The thick dashed line is the phase change when probe level is increased to 100 dB SPL (i.e., the difference between the thick dashed and the thick solid line in Fig. 1c). The thin dashed line is the phase change in the presence of a 8 kHz suppressor at 60 dB (thus, the difference between the dashed and the solid line in Fig. 2c). To the left of $x = 6.5$ [mm] the suppressed phase response follows approximately the thick dashed line since at these locations the effective BM impedance has been suppressed considerable [see Fig. 2(a)]. At locations to the right of $x = 7.2$ [mm] the suppressor does not affect the effective BM impedance, and the phase remains to the reference line. See text for the region between 6.9 and 7.2 [mm].

Thus, our simple long-wave model replicates the observed phase lags in the case of (non-dominant) HS suppression successfully. According to the attenuation hypothesis adding a suppressor to the stimulus has the same effect on the probe response as attenuation of probe level, so that phase leads should have been observed at locations basal to the probe peak and a transition at the peak [as can be inferred from Fig. 1(c)]. We conclude that the attenuation hypothesis fails in this example, and more examples will follow.

Can we explain the occurrence of phase lags at the probe peak? Yes, we can. An interesting property of our model is that the *course* of the phase (or magnitude) curves is locally related to the effective BM impedance, or (in good approximation) to the resistance component (see Appendix B). A consequence is that at places where the resistance is the same as in the passive (active) case the phase follows the passive (active) phase curve. Let us try to understand the phase course in Fig. 2(c) by looking at the resistance component in part (a) of the figure. We see the active mechanism has been suppressed considerably in the region basal to $x = 6.5$ [mm]. Accordingly, in part (c) the suppressed phase pattern follows at these location more or less the passive phase pattern [which is approximately the thick dashed line in Fig. 1(c)]. This is shown more clearly in Fig. 3 where phase changes have been plotted relative to the phase response of the 7 kHz probe at 30 dB SPL [thick solid line in Fig. 1(c)]. The thick dashed line indicates the relative phase change for the 7 kHz probe response at 100 dB SPL (from Fig. 1) which should be compared with the curves in Fig. 4 of Cheatham and Dallos (1989) and in Fig. 11(c) from Cheatham and Dallos (1990). The thin dashed line denotes the relative phase change of the suppressed probe in Fig. 2. We see that the thin and thick dashed lines more or less follow the same course at locations basal to $x = 6.5$ [mm]. Similarly, at locations apical to $x = 7.2$ [mm] the active mechanism is in full swing, and the thin dashed line in Fig. 3 consequently follows the reference line with a downward shift of about 90 degrees.[2] The magnitude of the response in Fig. 2(b) can be analysed similarly. A very nice illustration of our method is that in the region from $x = 6.9$ to 7.2 [mm] the phase lag increase in Fig. 3 can be seen to arise from the release of self-suppression in the same region in Fig. 2(a).

The effect of HS suppression on the probe changes when, at the location of the unsuppressed probe peak at low input levels, the suppressor becomes *larger* than the (suppressed) probe. This will be called *dominant* suppression. An example is given in Fig. 4 where we have decreased the suppressor frequency to 7.25 kHz. It is seen in Fig. 4(a) that the active mechanism near the probe peak has been suppressed more than in Fig. 2(a), while around $x = 5.5$ [mm] in Fig. 2 the active process has been nearly restored. Still, the response to the probe tone is suppressed much more in Fig. 4 than in Fig. 2 because the influence of the resistance on the probe response is largest near resonance (which lies approximately at the same location as the peak, at least for an input level not exceeding 30 dB SPL). Furthermore, we notice that the phase slope in Fig. 4(c) approaches the passive curve from Fig. 1(c) at those locations, basal to $x = 7.2$ [mm], where the active mechanism has been suppressed considerably. Apically to this point the phase curve follows the active curve. Because the suppressed phase curve crosses the unsuppressed curve with a smaller (in fact, less negative) slope than the unsuppressed curve, a phase lead (lag) arises at locations apical (basal) to this crossing point. Similar lag/lead behavior has been found in data on excitatory HS suppression in inner hair cells in the apical turn (Cheatham and Dallos, 1989, Figs. 7 and 8). A phase lead was also observed during dominant HS suppression for probe frequencies above CF by Nuttall and Dolan (1993b, Fig. 1, curve obtained for 18 kHz probe and 18.5 kHz suppressor). They suggested that a different type of suppression should exist for non-dominant HS suppression than for dominant HS suppression. The model results presented in Figs. 1 to 4 show that this is not necessary.

Another example of dominant HS suppression can be obtained when an 8 kHz suppressor (as in Fig. 2) is increased in level to 80 dB SPL. In that case phase lags still occur at all locations. When suppressor level is increased to 90 dB SPL, the phase behaviour becomes equivalent to that in Fig. 4(c), and the phase lags at and apically to the peak disappear. Similarly, Deng and Geisler (1985) observed in physiological responses that phase lags at CF disappeared when they increased HS suppressor level.

An example of non-dominant low-side (LS) suppression is shown in Fig. 5 where the suppressor frequency has been decreased to 5.5 kHz. We observe that the BM resistance at the frequency of the probe has now been increased mostly at and apically to

---

[2] Cheatham and Dallos (1990, 1993) stated that the observed phase lags could be explained by assuming that at locations basal to the peak, the velocity of the traveling wave (proportional to the reciprocal of the derivative of phase to frequency) decreases in the presence of a suppressor, which is true for our Fig. 2(c). However, they did not explain exactly how this velocity decrease was brought about.

the probe peak. The result is a decrease of the probe response peak, a decrease of the apical slope, and a phase lead at locations apical to the peak (except at $x \geq 8.7$ [mm] where the response will be too small to be measured). The occurrence of phase leads in this region is consistent with mechanical and inner hair cell data [Nuttall and Dolan, 1993a, Fig. 6(f) and 7(d); Cheatham and Dallos, 1990, Figs. 3(a) and 8(a)] but inconsistent with the attenuation hypothesis. The lag/lead transition becomes more evident when the suppressor is increased in level (not shown). The probe tone has been suppressed by 5 dB which is about the same as the values found by Ruggero *et al.* (1992, Fig. 4). Note that they used a different definition for suppression so that the suppression values in their Figs. 5 and 6 are larger than the actual decrease of the probe response peaks.

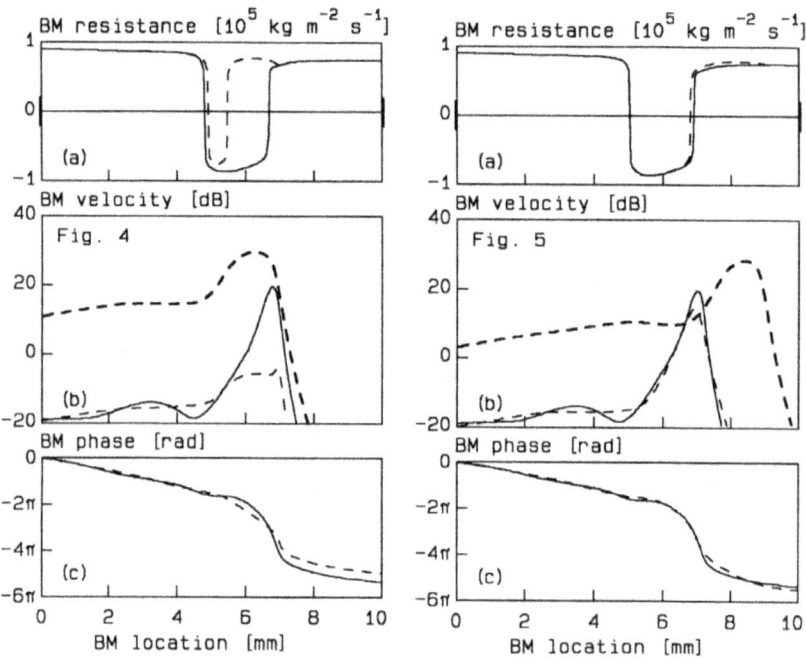

Figure 4. Dominant high-side suppression. (Dominant: at the unsuppressed probe peak the suppressor has a larger amplitude than the suppressed probe). The layout and parameter values are the same as in Fig. 2, but suppressor frequency is now 7.25 kHz. Note that at locations apical to the peak phase leads occur in contrast to the phase lags in Fig. 2.

Figure 5. Non-dominant low-side suppression. A 7 kHz tone at 30 dB SPL is suppressed by a 5.5 kHz tone at 55 dB SPL. Layout is the same as in Figs. 2 and 4.

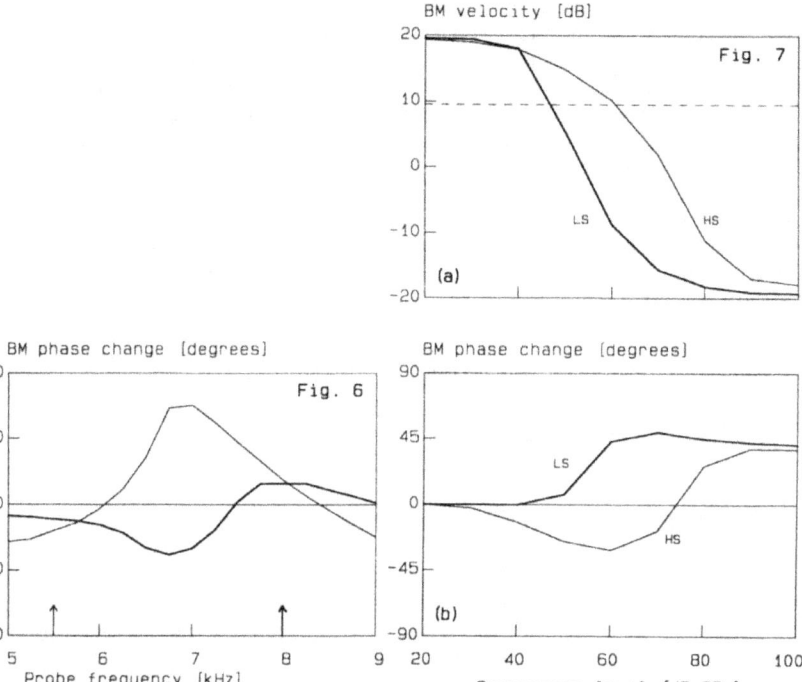

Figure 6. BM phase change in the presence of a fixed suppressor as a function of probe frequency. The thin line is the phase change of the probe response for a fixed suppressor at 5.5 kHz (thin arrow) and the thick line for a suppressor at 8 kHz (thick arrow). Probe at 70 dB SPL, suppressor at 80 dB SPL. Reference: unsuppressed probe response. Place of measurement: at x = 7.0 [mm] which is the location of the peak of a 7 kHz tone at 30 dB SPL.

Figure 7. Probe response as a function of suppressor level. Magnitude [part (a)] and phase shifts [part (b)] of a 7 kHz probe tone (at 30 dB SPL) in the presence of an HS (8 kHz) or LS (5.5 kHz) suppressor. The phase shifts are relative to the phase at a suppressor level of 20 dB SPL, and the place of measurement corresponds to the peak location of a 7 kHz probe tone at 30 dB SPL. The growth of suppression is much larger for the LS curve than for the HS curve when the amount of suppression is smaller than about 10 dB. For larger suppression values the growth of suppression are approximately equal. The phase shifts in part (b) for the HS case display a transition from lag to lead as suppressor level is increased.

We have considered a cochlea model that is place-frequency invariant (see Appendix A), so that the response patterns in our Fig. 1 can be compared isomorphically with iso-intensity response curves obtained as a function of frequency. This invariance property,

43

however, only holds for pure tones or for two-tone stimulation where the ratio of primary frequencies is kept constant. If the primary frequency ratio is not held constant, as in Fig. 2 and 4, phase behavior changes dramatically. Thus, we have to be careful to use the invariance property in relation with two-tone stimuli consisting of a fixed-frequency suppressor and a probe with varying frequency as has, for instance, been done by Cheatham and Dallos (1989, 1990). To make a fair comparison with the experimental data we produced Fig. 6 in which we varied probe frequency while suppressor frequency was kept at 5.5 kHz (thick line) or at 8 kHz (thin line). This figure shows changes of probe phase with respect to the unsuppressed condition. The curves have been obtained at the location where a 7 kHz tone has its peak at 30 dB SPL. The two curves show a lag/lead behavior with different transition frequencies. For probe frequencies approaching the (above-CF) suppressor frequency from below (see thick line) the suppressor becomes dominant and consequently a phase lead occurs (as in Fig. 4). This phase lead and the phase lags at lower probe frequencies are consistent with findings of Nuttall and Dolan (1993b, Fig. 2B) and Cheatham and Dallos (1989, Fig. 7). It is, however, inconsistent with data of Cheatham and Dallos (1989, Fig. 11; 1990, Figs. 3a and 5b), presumably because of the high levels used in their experiments. The thin line is similar to the curves in Figs. 3(b) and 5(b) of Cheatham and Dallos (1990).

3.3.2 Suppression and self-suppression: Intensity functions

If one plots the magnitude of the response at one location in the cochlea as a function of input level, an input-output (I/O) function is obtained. For linear responses the I/O function has a slope of 1 dB/dB. In our model where responses suffer from self-suppression (see Fig. 1), the minimal I/O slope for a pure tone is about 0.25 dB/dB (Kanis and de Boer, 1993b). In the presence of a suppressor, experimental I/O functions are shifted horizontally towards higher probe levels and, in addition, the I/O slope in the compressed region has often been found to increase (Abbas and Sachs, 1976; Javel et al., 1978; Robles et al., 1986b; Sokolowski et al., 1989; Ruggero et al., 1992; Rhode and Cooper, 1993; Nuttall and Dolan, 1993a). Thus, *self-suppression decreases the I/O slope of the probe in the nonlinear region, and suppression increases the slope.* One could also say that *suppression weakens self-suppression.* This linearization effect, as it is called, is also present in our model (not shown) and is brought about as follows. When there is no suppression, but only self-suppression, the minimal I/O slope of the probe is about 0.25 dB/dB. When the probe is fully suppressed, the nonlinear element is effectively put out of function, and the I/O slope becomes linear. For intermediate amounts of suppression the minimal slope of the I/O curve will thus be between 0.25 and 1 dB/dB. Note that this linearization effect is inconsistent with the attenuation hypothesis, since the attenuation hypothesis merely predicts a parallel shift of the I/O function towards higher intensities.

A different type of intensity function is obtained when not the probe level is varied but suppressor level. In neural, inner hair cell, cochlear microphonic and BM recordings it has been found that suppression increases with increasing suppressor level and that the growth of suppression is larger for LS suppressors than for HS suppressors (Abbas and Sachs, 1976; Cheatham and Dallos, 1982; Javel et al., 1983; Fahey and Allen, 1985; Costalupes et al., 1987; Delgutte, 1990; Ruggero et al., 1992; Rhode and Cooper, 1993;

Nuttall and Dolan, 1993a). These findings are replicated by the model results in Fig. 7 in which the range of suppression is fully covered (from no suppression at a suppressor level of 20 dB SPL to about 40 dB of suppression at a suppressor level above 90 dB SPL). This figure shows how magnitude and phase of a probe response at 30 dB SPL monitored at its peak are influenced by an LS or HS suppressor. The dashed line indicates where the probe response has been suppressed by an amount of about 10 dB with respect to the probe-alone case. While for both HS and LS suppressors the maximum amount of suppression is almost equal (being limited by the maximal enhancement of about 40 dB that can be produced by the active process), the suppression slopes are quite different for the two suppressors. For the HS suppressor the magnitude curve has a slope of about 0.4 dB/dB during the first 10 dB of suppression of the probe, while the slope for the LS suppressor has a value of about 1.3 dB/dB. Similar values have been found in the above-cited experiments, where it should be noted that in these papers different definitions of suppression have been used which makes a quantitative comparison difficult. The reason for the different slopes is that the influence of the resistance on the response is largest near resonance: for the LS suppressor the region where the resistance component is affected most lies at and apical to the probe peak [cf. Fig. 5(a)] whereas for the HS suppressor it lies mainly basally to the peak [see Fig. 2(a)]. All this applies to the case when there is less than 10 dB of suppression. For larger amounts of suppression the suppression slope becomes equal, or slightly larger, for the HS than for the LS suppressor. The reason for the changing of roles is that the influence of the HS suppressor on the probe has spread to locations around the probe peak, while for the LS suppressor the spread has been more to locations basal to the peak. Note that some of the intensity curves reported by Rhode and Cooper (1993) resemble our curves as a whole. Only these curves do not cover the whole range of about 40 dB suppression since they used a (self-suppressed) probe at 54 and at 74 dB SPL. (See also Fig. 5 of Nuttall and Dolan, 1990). Thus, a more detailed comparison of Fig. 7(a) with BM experiments must abide its time.

The phase leads in Fig. 7(b) for LS suppression are consistent with experimental data (Nuttall and Dolan, 1993a, Fig. 7d). As has already been mentioned, lag/lead phase behaviour during high-level LS suppression equals the phase behaviour in Fig. 1(c) when probe level is increased. Since the transition point in Fig. 1(c) is located 0.1 mm basal to the probe peak at 30 dB SPL, it is conceivable that during LS suppression phase leads are recorded if the phase is being monitored at the response peak, .

In Fig. 7(b) we observe for the HS case a transition from lag to lead as suppressor level is increased. If one extrapolates the trend in the phase shifts for low probe levels in Fig. 7(b) of Nuttall and Dolan (1993a) towards higher suppressor levels (above 70 dB SPL), a similar transition from lag to lead seems to occur in mechanical data. For the inner hair cell data shown in their Fig. 6(c) the slopes of the phase curves do not tend to go up as suppressor level is increased. A probable reason is that the ratio of suppressor to probe frequency is larger in their Fig. 6. Deng and Geisler (1985) found, also in line with our results, that when the phase of a probe tone at CF was monitored the observed phase lags disappeared when suppressor was increased.

3.4 Discussion

The purpose of this paper has been to clarify the mechanism of two-tone suppression with a nonlinear and locally active cochlea model. The model was solved in the frequency domain with the quasilinear method developed by us (Kanis and de Boer, 1993b). A great advantage of this solution method compared to the time-domain approach is that effects of saturation of the OHC pressure on the response can easily be monitored via the effective BM impedance. Another advantage is that the method is very fast (Kanis and de Boer, 1993b). Any response pattern as reported here is computed in less than one minute on a personal computer (486, 33 MHz).

We have shown that the phase behavior of the probe response is more complex during two-tone suppression than during probe level attenuation in that it depends critically on suppressor parameters. Phase behavior is not predicted as simply as is suggested by the (phenomenological) attenuation hypothesis. This hypothesis states that adding a suppressor to the input stimulus is equivalent to attenuation of the probe. For instance, a phase lag should always occur for probe frequencies higher than CF when the probe response at CF decreases in magnitude *irrespective of the cause of the magnitude decrease*.

In contrast, in our model *response changes at a certain location depend upon the dynamic state of the whole cochlea*. The dynamic state, in turn, is influenced by the presence of a suppressor. In other words, probe amplification at the peak location is reduced when the active mechanism is brought into saturation at other locations.

A property of our cochlea model is that the *slope of the phase curve* is *locally* related to the resistance component of the BM impedance. This relation enables us to understand, for instance, the occurrence of phase lags during non-dominant HS suppression.

As we have said earlier the phase behavior of the probe response depends critically on stimulus conditions; for instance, a small increase in HS-suppressor level can change the phase shift from negative to positive [see Fig. 7(b)]. Taking this into account, we may understand why some experimenters have had difficulty in explaining their phase data.

Other authors have described nonlinear models. Kim *et al.* (1973) and Hall (1974, 1977, 1980) solved passive models in the time domain. They were not able to simulate *non-dominant* LS suppression on the level of the BM. Hall could only simulate non-dominant LS suppression on the neural level by means of a second filter. Zwicker (1979, 1986) used a hardware model with a saturating feedback that *undamped* the system, that is, at all places less energy is injected by the active subsystem than absorbed (this is in contrast to *locally active* models where at certain places more energy is injected than absorbed). With his model Zwicker (1986) could not replicate non-dominant LS suppression. Locally active models are more powerful in that respect. Cohen and Furst (1993) have given an example of mutual (non-dominant) two-tone suppression, where they used an *ad hoc* activity distribution. Models similar to the one used in this paper were recently presented by Neely and Stover (1993) and Geisler *et al.* (1993). However, these authors did not systematically study two-tone suppression. Furthermore, these models were solved in the time domain.

Acknowledgements

This work was supported by the Netherlands Foundation for Scientific Research, project number 810-410-10-1. The authors thank three anonymous reviewers for their valuable comments.

Appendix 3.A: Model formulation

This section (which can be skipped by the non-mathematical reader) is the formal counterpart of section I. The linear long-wave model is described by the wave equation

$$p_{xx}(x;\omega) + k^2(x;\omega)\, p(x;\omega) = 0, \tag{3.A.1}$$

where $p(x;\omega)$ is the complex variable that denotes the pressure in one of the scalae and ω the radian frequency. The subscript xx stands for the second derivative with respect to x. The coefficient $k(x;\omega)$ has the dimension of a wave number, and is defined by

$$k^2(x;\omega) \equiv \frac{-2i\omega\rho}{hZ_{BM}(x;\omega)}, \tag{3.A.2}$$

in which ρ is the density of the fluid, h the effective height of the scalae, and $Z_{BM}(x;\omega)$ the impedance function that describes the BM. For a passive model we take $Z_{BM}(x;\omega)$ equal to

$$Z_{BM}^p(x;\omega) = M\omega_{loc}(x)\left[\delta + i(\beta(x;\omega) - 1/\beta(x;\omega))\right], \tag{3.A.3}$$

with $M = 0.5\,[\text{kg m}^{-2}]$, $\delta = 0.4$, and where $\beta(x;\omega)$ is defined as

$$\beta(x;\omega) \equiv \omega / \omega_{loc}(x). \tag{3.A.4}$$

The local resonance radian frequency of the BM, $\omega_{loc}(x)$, has the form:

$$\omega_{loc}(x) = \sqrt{\frac{S_0}{M}}\, \exp(-\alpha x / 2), \tag{3.A.5}$$

with $S_0 = 10^{10}\,[\text{kg m}^{-2}\,\text{s}^{-2}]$ and $\alpha = 3 \times 10^2\,[\text{m}^{-1}]$.

The cochlea is coupled to the pressure source in such a way that reflections of retrograde waves at the stapes are minimized.[3] For the coupling impedance, Z_0, we have taken an approximate form of the local impedance function (de Boer, 1991, section 5.6). The equations describing the coupling are

$$gp_0 - p(0;\omega) = - Z_0 v_{stapes},\qquad(3.A.6)$$

and

$$v_{stapes} = - p_x(0;\omega) / (i\omega\rho),\qquad(3.A.7)$$

where g is a scaling factor, p_0 the given input pressure and v_{stapes} the stapes velocity. For the frequencies used in this paper the coupling impedance is approximately constant and real. Equations (3.A.6) and (3.A.7) form the boundary condition for the model at $x = 0$. The boundary condition at the helicotrema (at $x = x_{end}$) is described by

$$p(x_{end};\omega) = 0.\qquad(3.A.8)$$

Once the wave equation has been solved, the BM velocity $v(x;\omega)$ is calculated from

$$v(x;\omega) = - 2p(x;\omega) / Z_{BM}(x;\omega).\qquad(3.A.9)$$

In all computations the helicotrema has been set equal to 1 cm (since only high frequencies are used) and the number of sections is 500.

We assume that the outer hair cells (OHCs) generate a pressure $P_{OHC}(x;\omega)$ that is added (somehow) to the pressure difference across the BM. If we write this pressure as

$$P_{OHC}(x;\omega) = Z_{OHC}(x;\omega)\, v(x;\omega),\qquad(3.A.10)$$

the passive BM impedance $Z^p_{BM}(x;\omega)$ is modified by the feedback to the locally active BM impedance $Z^a_{BM}(x;\omega)$ according to

$$Z^a_{BM}(x;\omega) = Z^p_{BM}(x;\omega) - Z_{OHC}(x;\omega).\qquad(3.A.11)$$

The transfer impedance $Z_{OHC}(x;\omega)$ is the same as the one used in Kanis and de Boer (1993a, 1993b). It contains a factor $G(x;\omega)$ that describes the filtering of the BM velocity by a resonance of the tectorial membrane and the stereocilia of the OHCs (Allen, 1980; Neely and Kim, 1986) followed by another filtering, and a factor $e_0\omega_{loc}(x)$ due to electro-mechanical transduction:

---

[3] In fact, this way of "smooth" coupling is only necessary for the computation of distortion products, since for the primary components the retrograde waves are 40 dB smaller than the anterograde waves. We use it here to maintain consistency between different papers (Kanis and de Boer, 1993a, 1993b, 1994).

$$Z_{OHC}(x;\omega) = e_0 \, w_{loc}(x) \, G(x;\omega). \tag{3.A.12}$$

In our model we have

$$G(x;\omega) = d_0 \frac{1 + i\beta(x;\omega)}{\delta_{SC} + i[\beta(x;\omega) - \sigma^2 / \beta(x;\omega)]}. \tag{3.A.13}$$

The numerical values of the parameters are $d_0 = 1404$ [kg s^{-1}], $e_0 = 4.28 \times 10^{-5}$ [kg m^{-2} s], $\delta_{SC} = 0.14$ and $\sigma = 0.7$. With this activity distribution, a 40 dB enhancement of the velocity response is achieved, and the form of the BM-response peak is compatible with mechanical data (cf. Sellick et al., 1982; Robles et al., 1986a). The Q_{10} value of about 6.5 is independent of frequency in our model (but it is, of course, dependent on level). We have checked the stability of the model by looking at the resistance part of the input impedance at the stapes which had to be positive for all frequencies. The value of the coefficient g in Eq. (3.A.4) has been given a value of 3.5×10^2 so that the BM-response peak of the active model is scaled to the response measured by Sellick et al. (1982, Fig. 15, curve with closed circles).

Since the OHC potential is found to be saturating (Hudspeth and Corey, 1977), we have assumed a linear transformation of cell-potential into pressure and have placed the nonlinearity between $G(x;\omega)$ and the electro-mechanical transduction. The generation of the nonlinear OHC pressure $P^{NL}_{OHC}(x,t)$ is described by

$$P^{NL}_{OHC}(x,t) = e_0 w_{loc}(x) \tanh(I(x,t)), \tag{3.A.14}$$

where $I(x,t)$ is a real variable that represents the actual waveform of the input to the OHCs. In the case of two-tone stimulation, the input to the nonlinearity consists of two filtered primary components of the BM-velocity besides many (filtered) distortion components. Since the distortion components are small and have negligible influence on the primary components of the BM-velocity (see Kanis and de Boer, 1993b, Appendix B) we can write $I(x,t)$ as consisting of only two components:

$$I(x,t) = \sum_{k=1,2} |G(x;\omega_k) \, v(x;\omega_k)| \sin(\omega_k t + \varphi(x;\omega_k)), \tag{3.A.15}$$

where ω_k denotes the radian frequency of component k, and $\varphi(x;\omega_k)$ the phase of $G(x;\omega_k)v(x;\omega_k)$. The complex variable $v(x;\omega_k)$ denotes the first-order velocity component with radian frequency ω_k; it is defined by

$$v(x;\omega_k) = -2i \int_0^T \frac{dt}{T} \, v(x,t) \exp(-i\omega_k t), \tag{3.A.16}$$

49

where T is equal to the period of repetition of the waveform of the BM velocity $v(x,t)$.

We want to create a linear relation between pressure and velocity so that the familiar concept of transfer impedance can be used. Thus, instead of the nonlinear OHC pressure $p^{NL}_{OHC}(x,t)$ of Eq. (3.A.14) we use its first Fourier coefficient; it is denoted by $P_{OHC}(x;\omega_k)$ and given by

$$P_{OHC}(x;\omega_k) = e_0\omega_{loc}(x) \int_0^T \frac{dt}{T} \tanh(I(x,t)) \exp(-i\omega_k t). \tag{3.A.17}$$

Dividing $P_{OHC}(x;\omega_k)$ from Eq. (3.A.17) by $v(x;\omega_k)$ gives us the quasilinear OHC impedance $Z^{QL}_{OHC}(x;\omega_k)$ for the coefficient with frequency $\omega_k/2\pi$. The thus-obtained impedance is used to modify the passive BM impedance as follows:

$$Z^{QL}_{BM}(x;\omega_k) = Z^P_{BM}(x;\omega_k) - Z^{QL}_{OHC}(x;\omega_k). \tag{3.A.18}$$

The essence of the quasilinear method is to use $v(x;\omega_k)$ and $Z^{QL}_{BM}(x;\omega_k)$ in a linear frequency-domain model. First, eight iteration steps are used to calculate the quasilinear response of one component alone and eight for the other component alone. Then, two steps are used for each component to compute the effect of suppression by the other component. It is found that the effect of adding higher-order terms in Eq. (3.A.15) on the computed BM responses is negligibly small. The reason is that the distortion products are not large enough to be able to suppress the primaries. Thus, omitting them in Eq. (3.A.15) is justified. For details on the computation of distortion products we refer to Kanis and de Boer (1993a, 1993b, 1994).

It should be noted that the value of d_0 has been chosen in such a way that the onset of nonlinearity of a pure tone starts at about the same input level (30 dB SPL) as in experiments. From Eq. (3.A.14) we see that $e_0\omega_{loc}(0)$ is the maximum pressure generated by the OHCs. It has a value of about 6 N m^{-2} (which follows from our model by dividing $e_0\omega_{loc}(0)d_0$ by d_0). Assuming a BM width of 300 μm and an OHC width of 10 μm, the maximum force for one OHC is about 2 nN. Notice that this value is comparable to the force that can be expected to be produced by a OHC *in situ* (Iwasa and Chadwick, 1992).

Appendix 3.B: Relating phase changes to saturation of the effective BM impedance

In our cochlea model we need to know the dynamic state of the whole cochlea to compute the phase (or magnitude) of the response at a certain location. But to compute the phase (or magnitude) *derivative* at that location we need to know the effective impedance only at that location. This is proved as follows. In a long-wave model the BM response is described reasonably well by the LG (Liouville-Green) or WKBJ (Wentzel-Kramers-Brillouin-Jeffreys) solution (Zweig *et al.*, 1976):

$$v(x;\omega) = c(\omega)k^{3/2}(x;\omega)\, e^{-i \int_0^x k(x';\omega)\, dx'}, \tag{3.B.1}$$

or

$$\ln(v(x;\omega)) = \ln(c(\omega)) + 3/2\,\ln(k(x;\omega)) - i \int_0^x k(x';\omega)\, dx'. \tag{3.B.2}$$

The coefficient $c(\omega)$ is equal to $v(0;\omega)k^{-3/2}(0;\omega)$. The coefficient $k(x;\omega)$ depends on the effective BM impedance [Eq. (3.A.18)] as described in Eq. (3.A.2). It should be noted that Eq. (3.B.1) is only true when there is no reflection.[4] For an input level of 30 dB SPL the 7 kHz probe in Fig. 1 displays some reflection. Also, in the presence of a suppressor, for instance, the BM impedance might become distorted giving rise to reflections. However, in our model results Eq. (3.B.1) remains a good approximation to the solution of Eq. (3.A.1). The derivative of the phase with respect to location becomes:

$$\frac{\partial \varphi_v(x;\omega)}{\partial x} = -\,\mathrm{Re}\{k(x;\omega)\} - 3/4\,\frac{\partial \varphi_Z(x;\omega)}{\partial x}, \tag{3.B.3}$$

where $\varphi_v(x;\omega)$ and $\varphi_Z(x;\omega)$ are the phases of the velocity $v(x;\omega)$ and $Z_{\mathrm{BM,QL}}(x;\omega)$, respectively. The integral over $k(x;\omega)$ has disappeared, so that $\partial \varphi_v(x;\omega)/\partial x$ depends only *locally* on the effective impedance at location x.[5] Thus, although the phase is a global function of the effective impedance function, the phase derivative depends *locally* on the impedance. Furthermore, the variations of the phase patterns can be predicted reasonably well by looking at the *resistance* component alone. The reactance component may be ignored since variations of the reactance component in the region of suppression are reflected in the resistance. This means that at places where the resistance component of the effective BM impedance is the same as in the passive case the phase follows the passive phase curve, and where the feedback mechanism operates in the linear regime, i.e., where it is not saturated, the phase curve follows the phase curve belonging to the active case. We use this knowledge to understand phase shifts in the model results presented in this paper.

---

[4] The LG approximation also breaks down in the basal turn for very low frequencies; in that case Hankel instead of exponential functions should be used (Zwislocki, 1948; Shera and Zweig, 1991).

[5] Note that the effective impedance and, consequently, the phase derivative may still be a global function of x. Equation (B3) only states that there is a *one-to-one* relation between the phase derivative and the quasilinear impedance at location x.

References

Abbas, P.J., and Sachs, M.B. (1976). "Two-tone suppression in auditory nerve fibers: Extension of a stimulus-response relationships," J. Acoust. Soc. Am. 59, 112-122.

Allen, J.B. (1980). "Cochlear micromechanics—A physical model of transduction," J. Acoust. Soc. Am. 68, 1660-1679.

Anderson, D.J., Rose, J.E., Hind, J.E., and Brugge, J.F. (1971). "Temporal position of discharge in single auditory nerve fibers within the cycle of a sine-wave stimulus: Frequency and intensity effects, " J. Acoust. Soc. Am. 49, 1131-1139.

Arthur, R.M., Pfeiffer, R.R., and Suga, N. (1971). "Properties of two-tone inhibition in primary auditory neurons," J. Physiol. 212, 593-609.

Boer, E. de (1991). "Auditory Physics. Physical principles in hearing theory. III," Phys. Rep. 203, 125-231.

Cheatham, M.A., and Dallos, P. (1982). "Two-tone interaction in the cochlear microphonic," Hear. Res. 8, 29-48.

Cheatham, M.A., and Dallos, P. (1989). "Two-tone suppression in inner hair cell responses," Hear. Res. 40, 187-196.

Cheatham, M.A., and Dallos, P. (1990). "Comparison of low- and highside two-tone suppression in inner hair cell and organ of Corti responses," Hear. Res. 50, 193-210.

Cheatham, M.A., and Dallos, P. (1993). "Comment on 'Two-tone suppression of inner hair cell and basilar membrane responses in the guinea pig,' " J. Acoust. Soc. Am. 94, 3509-3510.

Cohen, A., and Furst, M. (1993). "Cochlear model for rate suppression based on cochlear amplifier dynamics," in: in Biophysics of Hair-cell Systems, edited by H. Duifhuis, J.W. Horst, P. van Dijk and S.M. van Netten (World Scientific, Singapore), pp. 323-329.

Costalupes, J.A., Rich, N.C., and Ruggero, M.A. (1987). "Effects of excitatory and non-excitatory suppressor tones on two-tone rate suppression in auditory nerve fibers," Hear. Res. 26, 155-164.

Covell, W.P., and Black, L.J. (1936). "The cochlear response as an index of hearing," Am. J. Physiol. 116, 524-530.

Dallos, P. (1986). "Neurobiology of cochlear inner and outer hair cells: Intracellular recordings," Hear. Res. 22, 185-198.

Delgutte, B. (1990). "Two-tone rate suppression in auditory-nerve fibers: Dependence on suppressor frequency and level," Hear. Res. 49, 225-246.

Deng, L., and Geisler, C.D. (1985). "Changes in the phase of excitor-tone responses in cat auditory-nerve fibers by suppressor tones and fatigue," J. Acoust. Soc. Am. 78, 1633-1643.

Fahey, P.F., and Allen, J.B. (1985). "Nonlinear phenomena as observed in the ear canal and at the auditory nerve," J. Acoust. Soc. Am. 77, 599-612.

Geisler, C.D., and Sinex, D.G. (1980). "Responses of primary auditory fibers to combined noise and tonal stimuli," Hear. Res. 3, 317-334.

Geisler, C.D., Yates, G.K., Patuzzi, R.B., Johnstone, B.M. (1990). "Saturation of outer hair cell receptor currents causes two-tone suppression," Hear. Res. 44, 241-256.

Geisler, C.D., Bendre, A., and Liotopoulos, F.K. (1993). "Time-domain modeling of a nonlinear, active model of the cochlea," in *Biophysics of Hair-cell Systems*, edited by H. Duifhuis, J.W. Horst, P. van Dijk and S.M. van Netten (World Scientific, Singapore), pp. 330-337.

Hall, J.L. (1974). "Two-tone distortion products in a nonlinear model of the basilar membrane," J. Acoust. Soc. Am. 56, 1818-1828.

Hall, J.L. (1977). "Two-tone suppression in a nonlinear model of the basilar membrane," J. Acoust. Soc. Am. 61, 802-810.

Hall, J.L. (1980). "Cochlear models: Two-tone suppression and the second filter," J. Acoust. Soc. Am. 67, 1722-1728.

Hudspeth, A.J., and Corey, D.P. (1977). "Sensitivity, polarity, and conductance change in the response of vertebrate hair cells to controlled mechanical stimuli," Proc. Natl. Acad. Sci. USA 74, 2407-2411.

Iwasa, K.H., and Chadwick, R.S. (1992). "Elasticity and active force generation of cochlear outer hair cells," J. Acoust. Soc. Am. 92, 3169-3173.

Javel, E., Geisler, C.D., and Ravindran, A. (1978). "Two-tone suppression in the auditory nerve of the cat: Rate-intensity and temporal analysis," J. Acoust. Soc. Am. 63, 1093-1104.

Javel, E., McGee, J., Walsh, E.J., Farley, G.R., and Gorga, M.P. (1983). "Suppression of auditory nerve responses. II. Suppression threshold and growth, iso-suppression contours," J. Acoust. Soc. Am. 74, 801-813.

Kanis, L.J., and Boer, E. de (1993a). "The emperor's new clothes: DP emissions in a locally-active nonlinear model of the cochlea," in *Biophysics of Hair Cell Sensory Systems*, edited by H. Duifhuis, J.W. Horst, P. van Dijk and S.M. van Netten (World Scientific, Singapore), pp. 304-311.

Kanis, L.J., and Boer, E. de (1993b). "Self-suppression in a locally active nonlinear model of the cochlea: A quasilinear approach," J. Acoust. Soc. Am. 94, 3199-3206.

Kanis, L.J., and Boer, E. de (1994). "Frequency dependence of acoustic distortion products in a locally active model of the cochlea," submitted to the Journal of the Acoustical Society of America.

Kim, D.O., Molnar, C.E., and Pfeiffer, R.R. (1973). "A system of nonlinear differential modeling basilar-membrane motion," J. Acoust. Soc. Am. 54, 1517-1529.

Kim, D.O. (1986). "A review of nonlinear and active cochlea models," in: *Peripheral Auditory Mechanisms*, edited by J.B. Allen, J.L. Hall, A. Hubbard, S.T. Neely and A. Tubis (Springer-Verlag, Berlin), pp. 239-249.

Mountain, D.C., Hubbard, A.E., and McMullen, T.A. (1983). "Electromechanical processes in the cochlea," in: *Mechanics of Hearing*, edited by E. de Boer and M.A. Viergever (Univ. Press, Delft), pp. 119-126.

Neely, S.T., and Kim, D.O. (1986). "A model for active elements in cochlear biomechanics," J. Acoust. Soc. Am. 79, 1472-1480.

Neely, S.T., and Stover, L.J. (1993). "Otoacoustic emissions from a nonlinear, active model of cochlear mechanics," in: *Biophysics of Hair Cell Sensory Systems*, edited by H. Duifhuis, J.W. Horst, P. van Dijk, and S.M. van Netten (World Scientific, Singapore), pp. 64-70.

Nuttall, A.L., and Dolan, D.F. (1990). "Inner hair cell responses to the $2f_1$-f_2 intermodulation distortion product," J. Acoust. Soc. Am. 87, 782-790.

Nuttall, A.L., and Dolan, D.F. (1993a). "Two-tone suppression of inner hair cell and basilar membrane responses in the guinea pig," J. Acoust. Soc. Am. 93, 390-400.

Nuttall, A.L., and Dolan, D.F. (1993b). "Response to 'Comment on "Two-tone suppression of inner hair cells and basilar membrane responses in the guinea pig," ' " J. Acoust. Soc. Am. 94, 3511-3514.

Patuzzi, R.B., Yates, G.K., and Johnstone, B.M. (1989). "Outer hair cell receptor current and sensorineural hearing loss," Hear. Res. 42, 47-72.

Rhode, W.S. (1977). "Some observations on two-tone interaction measured with the Mössbauer effect," in *Psychophysics and Physiology of Hearing*," edited by E.F. Evans and J.P. Wilson (Academic, London), pp. 27-38.

Rhode, W.S., and Cooper, N.P. (1993). "Two-tone suppression and distortion production on the basilar membrane in the hook region of cat and guinea pig cochleae," Hear. Res. 66, 31-45.

Rhode, W.S., and Robles, L. (1974). "Evidence from Mössbauer experiments for nonlinear vibration in the cochlea," J. Acoust. Soc. Am. 55, 588-596.

Robles, L., Ruggero, M.A., and Rich, N.C. (1986a). "Basilar membrane mechanics at the base of the chinchilla cochlea. I. Input-output functions, tuning curves, and response phases," J. Acoust. Soc. Am. 80, 1364-1374.

Robles, L., Ruggero, M.A., and Rich, N.C. (1986b). "Mössbauer measurements of the mechanical response to single-tone and two-tone stimuli at the base of the chinchilla cochlea," in *Peripheral Auditory Mechanisms*, edited by J.B. Allen, J.L. Hall, A. Hubbard, S.T. Neely and A. Tubis (Springer-Verlag, Berlin), pp. 121-128.

Robles, L., Ruggero, M.A., and Rich, N.C. (1989). "Nonlinear interactions in the mechanical response of the cochlea to two-tone stimuli," in *Cochlear Mechanics - Structure, Function and Models*, edited by J.P. Wilson and D.T. Kemp (Plenum Press, New York), pp. 121-128.

Ruggero, M.A., Robles, L., and Rich, N.C. (1992). "Two-tone suppression in the basilar membrane of the cochlea: Mechanical basis of auditory-nerve rate suppression," J. Neurophysiol. 68, 1087-1099.

Sachs, M.B. (1969). "Stimulus-response relation for auditory-nerve fibers: Two-tone stimuli," J. Acoust. Soc. Am. 45, 1025-1036.

Sachs, M.B., and Abbas, P.J. (1976). "Phenomenological model for two-tone suppression," J. Acoust. Soc. Am. 60, 1157-1163.

Sachs, M.B., and Kiang, N.Y.-S. (1968). "Two-tone inhibition in auditory-nerve fibers," J. Acoust. Soc. Am. 43, 1120-1128.

Sellick, P.M., and Russell, I.J. (1979). "Two-tone suppression in cochlear hair cells," Hear. Res. 1, 227-236.

Sellick, P.M., Patuzzi, R., and Johnstone, B.M. (1982). "Measurement of basilar membrane motion in the guinea pig using the Mössbauer technique," J. Acoust. Soc. Am. 72, 131-141.

Shera, C.A., and Zweig, G. (1991). "Reflection of retrograde waves within the cochlea and at the stapes," J. Acoust. Soc. Am. 89, 1290-1305.

Sokolowski, B.H.A., Sachs, M.B., and Goldstein, J.L. (1989). "Auditory nerve rate-level functions for two-tone stimuli: Possible relation to basilar membrane nonlinearity," Hear. Res. 41, 115-124.

Zweig, G., Lipes, R. and Pierce, J.R. (1976). "The cochlear compromise," J. Acoust. Soc. Am. 59, 975-982.

Zweig, G. (1991). "Finding the impedance of the organ of Corti," J. Acoust. Soc. Am. 89, 1229-1254.

Zwicker, E. (1979). "A model describing nonlinearities in hearing by active processes with saturation at 40 dB," Biol. Cybernetics 35, 243-250.

Zwicker, E. (1986). "A hardware cochlear nonlinear preprocessing model with active feedback," J. Acoust. Soc. Am. 80, 146-153.

Zwislocki, J. (1948). "Theorie der Schneckenmechanik: Qualitative und quantitative Analyse," Acta Otolaryng. Suppl. 72.

4

Frequency dependence of acoustic distortion products in a locally active model of the cochlea*)

Abstract   In two-tone experiments it has been shown that acoustic distortion products are 'tuned' as a function of primary frequency ratio, that is, at a certain frequency ratio a maximum in emission occurs. Several authors maintain that this 'tuning' is caused by band-pass filtering of the distortion products as they are coupled back to the basilar membrane. This view is challenged in the current paper. It is shown that the same kind of 'tuning' is present in a cochlea model without such a filtering of distortion products. It is argued that the observed 'tuning' is caused by a nonlinear mechanism inside the cochlea situated at the location of the outer hair cells. This idea is supported by several model results, and suggestions are made for future experiments in which this idea is to be tested further.

4.1 Introduction

It has widely been accepted that outer hair cells (OHCs) are essential for the improvement of frequency selectivity and sensitivity of the ear. The exact mechanism of how pressures produced by the OHCs influence the motion of the basilar membrane is not known, but several micromechanical models of the cochlea have been devised with OHCs as the pressure (or velocity) generators (Geisler, 1991; Neely, 1993; Neely and Stover, 1993). These locally active models have in common that in every cochlear section the OHCs are imbedded in the organ of Corti. That is, the pressures generated by the OHCs have to work against the internal mechanics of the organ of Corti, so that only part of these pressures is available to amplify the travelling wave (see also de Boer, 1991). This embedding results in a *filtering* of the OHC-generated pressures as they are coupled back to the basilar membrane. Several authors have argued that this or a similar filtering mechanism can be detected in the behaviour of distortion products (DPs) in otoacoustic emissions when two tones with frequency f_1 and f_2 are used as stimulus

---

*) Preliminary results were presented during the *1995 Midwinter Meeting* of the Association for Research in Otolaryngology, St. Petersburg, USA.

57

(Brown and Gaskill, 1990a; Brown and Williams, 1993; Allen and Fahey, 1993). Their argumentation is based on the experimental fact that acoustic distortion products are (band-pass) 'tuned' as a function of primary frequency ratio f_2/f_1, i.e., maximal emissions occur at a ratio of about 1.2 (Kim, 1980; Wilson, 1980; Fahey and Allen, 1986; Lonsbury-Martin et al., 1987; Harris et al., 1988; Brown and Gaskill, 1990a, 1990b; Gaskill and Brown, 1990; Whitehead et al., 1992). The increase in DP emission when the frequency ratio is decreased from 1.5 to 1.2 is understandable if one bears in mind that the overlap between the two primary responses increases in that case. Then, the active mechanism is driven more into saturation, and more DP generation will occur. The overlap becomes complete when the ratio f_2/f_1 is reduced from 1.2 to 1.0, and DP generation becomes maximal. However, such a monotonic increase in DP emission has not been seen in the experimental data. On the contrary: DP emission decreases when the frequency ratio goes from 1.2 towards 1.0. This is why a filtering mechanism such as the one described above is held responsible for the observed DP 'tuning': the location where the DP is generated changes for varying primary frequency ratios so that the DP component in the OHC pressure is 'moved' through the filter. Neely and Stover (1993) have found similar 'tuning' of acoustic DPs in a time-domain implementation of their nonlinear cochlea model that contains DP filtering.

In this paper, however, the necessity to invoke DP filtering in order to explain the 'tuning' of acoustic DPs is questioned. It will be shown that similar 'tuning' can be achieved with a nonlinear locally active model in which there is no filtering of the DP component after it has been generated by the OHCs. (Of course, filtering remains involved in the input path to the OHCs.) In other words, one cannot conclude from the available experimental emission data that a DP filtering mechanism is present in the real cochlea. We will also show that the 'tuning' disappears at levels where our cochlea model operates more in the linear regime. Thus, we conjecture that the DP 'tuning' seen in acoustic emission data is the consequence of a nonlinearity in the active process and not the consequence of frequency-selective filtering.

4.2 Model and method

In macromechanical models of the cochlea the cochlear partition may be described by an impedance consisting of a mass, a resistance, and a stiffness part. A micromechanical model may be described by three impedances, for instance Z_{BM}, Z_{RL}, and Z_{OC}; the first describing the mechanics of the basilar membrane (BM), and the other two describing the mechanics of the reticular lamina (RL) and the organ of Corti (OC). This is shown schematically in Fig. 1, where $-2p$ denotes the (fluid) pressure difference over the cochlear partition and v_{BM} and v_{RL} stand for BM velocity and RL velocity, respectively. When the BM and RL move relative to each other the size of the organ of Corti is changed. The difference between v_{BM} and v_{RL} is the velocity of the OC, v_{OC}. Thus, in the case of Fig. 1, the total impedance of the cochlear partition consists of the BM impedance and a parallel combination formed by the impedances of the RL and the OC. In a different model the parallel combination might be formed by the impedance of the TM and the impedance that describes the stereociliar coupling with the TM.

In an active cochlea model the outer hair cells (OHCs) amplify the travelling wave by generating an extra pressure difference over the cochlear partition. Since the OHCs are embedded in the organ of Corti, the pressure source is to be added in series with the impedance of the organ of Corti (at the location of the open circle in Fig. 1), so that the OHC pressures are filtered as they are coupled back to the BM [see Eq. (4.A.5) and (4.A.6) of the Appendix]. Several known locally active cochlea models are of this kind (Geisler, 1991; Geisler *et al.*, 1993; Neely, 1993; Neely and Stover, 1993). In order to see in how far the 'tuning' is the result of the presence of a 'second filter' we used a model in which the OHC pressures were not filtered after they had been generated. The model we used was similar to that of Neely and Kim (1986). In their model the OHCs are not imbedded in the organ of Corti and the effective action of the OHCs is put directly over the cochlear partition (at the location of the filled circle in Fig. 1).

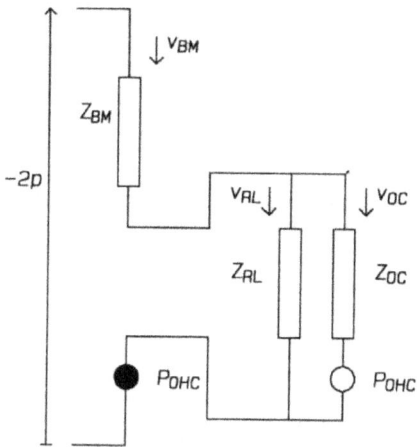

Figure 1. Electrical circuit for a cochlear section. The impedance Z_{RL} describes how the reticular lamina (RL) are attached to the modiolus via the tectorial membrane, Z_{OC} describes the internal mechanics of the organ of Corti, and Z_{BM} is the BM impedance. The velocity of the BM is denoted by v_{BM}, the velocity of the RL by v_{RL}, and the velocity of amplitude changes of the organ of Corti by v_{OC}. The pressure P_{OHC} is the pressure generated by the outer hair cells, and $-2p$ is the (fluid) pressure difference over the cochlear partition.

In the cochlea stimulation at the stapes sets up a travelling wave along the cochlear partition. When two primary components are present in the input stimulus, two peaks will occur in the velocity response of the cochlear partition with the more basal peak belonging to the higher frequency. If both primaries are strong, they will give rise to several DP components in the locally generated OHC pressure. In what follows DP will denote only the component with frequency $2f_1$-f_2. In our model the DP pressure component can, in every cochlear section, be viewed as a pressure source (see the filled circle in Fig. 1) that stimulates the BM directly without being filtered. These locally-produced DP pressure components generate a travelling wave at the frequency of the DP which during its travel towards the place of resonance will be amplified by the OHCs. Amplification occurs because the BM velocity contains components $v_{BM}(x;\omega_{DP})$ with the DP frequency. Thus, not only is a DP component generated at places where the pressure source is pushed into saturation, but the resultant travelling wave is also amplified by pressure sources at other locations. This means that the total input to the OHCs consists of both the two primary components and the DP component. The DP pressure component generated by the OHCs corresponding to this input will be called $P_{OHC}(x;\omega_{DP})$.

In this paper the response of the nonlinear model is solved in the frequency domain by considering only the relevant Fourier components in the pressures generated by the OHCs. Thus, we first compute the responses to the two primaries in a number of iterations as described in Kanis and de Boer (1994); then we proceed with the computation of the DP response. We assume that we may neglect the influence of the DP response on the primaries since, for single-tone stimulation, the influence of higher-order components on primary components is negligible (see Kanis and de Boer, 1993b, Appendix B). How do we solve for the DP response? With the two computed primary responses we compute $P^{(1)}_{OHC}(x;\omega_{DP})$ and $Z^{(1)}_{OHC}(x;\omega_{DP})$, initial guesses to the pressure distribution $P_{OHC}(x;\omega_{DP})$ and the active impedance at the DP frequency $Z_{OHC}(x;\omega_{DP})$, respectively. [If $Z_{OHC}(x;\omega_{DP})$ is not suppressed by the primaries it is of the form given by Eq. (9) in Kanis and de Boer (1993b).] These two distributions lead to the first estimate $v^{(1)}_{BM}(x;\omega_{DP})$ and $p^{(1)}(x;\omega_{DP})$ of the velocity response $v_{BM}(x;\omega_{DP})$ and pressure response $p(x;\omega_{DP})$, respectively, by solving a linear problem [see Kanis and de Boer, 1993b, Eq. (B7)]. In the second iteration step we compute $P^{(2)}_{OHC}(x;\omega_{DP})$ which is the local pressure generated by an OHC if both primaries and the first guess to the DP response are used as input to the OHC, so that $P^{(2)}_{OHC}(x;\omega_{DP})$ also contains amplification effects by the DP response itself. This is a better guess to the pressure distribution $P_{OHC}(x;\omega_{DP})$ than $P^{(1)}_{OHC}(x;\omega_{DP})$. The estimate for the impedance function to be used in the next iteration step is:

$$Z^{(k)}_{OHC}(x;\omega_{DP}) \equiv \frac{P^{(k)}_{OHC}(x;\omega_{DP}) - P^{(1)}_{OHC}(x;\omega_{DP})}{v^{(k-1)}_{BM}(x;\omega_{DP})}, \tag{4.1}$$

where k is the iteration index, $k > 1$. In theory the iteration sequence can be carried on indefinitely. In practice two iteration steps suffice to get a good estimate of $P_{OHC}(x;\omega_{DP})$, $Z_{OHC}(x;\omega_{DP})$, $v_{BM}(x;\omega_{DP})$, and $p(x;\omega_{DP})$. It should be noted that in all iteration steps we use $P^{(1)}_{OHC}(x;\omega_{DP})$ as the 'distributed excitation' to the cochlea. Details of the computational method are given in Kanis and de Boer (1993b) and Kanis and de Boer (1994, Appendix A).

An additional remark concerns the notion of virtual fluid mass riding with the cochlear partition (Neely, 1985, Appendix B). If this concept is implemented in our model, the pressure distribution changes only somewhat near the response peak, but the velocity response remains the same. Also, the pressure difference at the stapes computed for different primary frequency ratios (as has been done in this paper and in Kanis and de Boer, 1993a) is not affected since the fluid-mass term does not influence power dissipation. To account for the virtual mass we have chosen the BM mass as rather high.

The parameters of the model were equal to those used in Kanis and de Boer (1993b). To monitor the emission of the combination tone we have considered the pressure at the location of the stapes, outside the cochlea.

4.3 Results

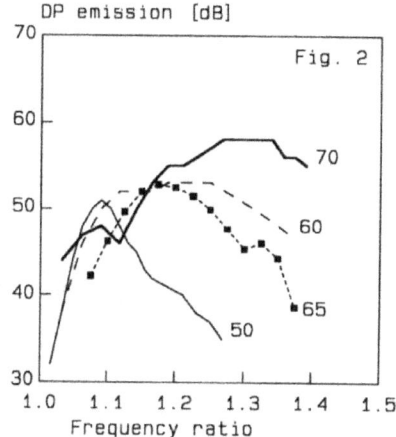
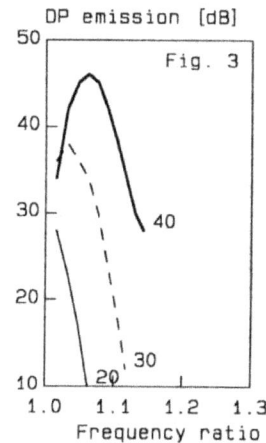

Figure 2. DP emissions as a function of primary frequency ratio displayed for three different primary input levels (50 dB SPL: thin solid line; 60 dB SPL dashed line; 70 dB SPL: thick solid line). We have monitored the pressure outside the cochlea at the location of the stapes with a reference level of 1×10^{-5} N m^{-2}. The experimental curve with the squares is taken from Gaskill and Brown (1990, Fig 2d, dashed line). The reference level of the experimental curve is 3.17×10^{-8} N m^{-2}.

Figure 3. DP emissions as a function of primary frequency ratio. Same as Fig. 2 but for lower input levels (20 dB SPL: thin solid line; 30 dB SPL dashed line; 40 dB SPL: thick solid line).. The tuning is less than at higher levels.

In Fig. 2 DP emissions computed with the model are shown as a function of primary frequency ratio where we have held the DP frequency constant. Results are displayed for three different primary input levels. To monitor the emission we have considered the pressure at the location of the stapes outside the cochlea. The curves show similar 'tuning' as the experimental curve with the squares taken from Gaskill and Brown (1990, Fig. 2d, dashed line). In Fig. 2 we have held the DP frequency constant. In many experiments f_2 is kept constant, but this gives essentially the same results. The peak in the 'tuning' curves in Fig. 2 shift towards higher ratios when primary levels are increased. This has also been found in experiments on acoustic DPs (Kim, 1980; Harris *et al.*, 1988; Lonsbury-Martin *et al.*, 1987; Gaskill and Brown, 1990).

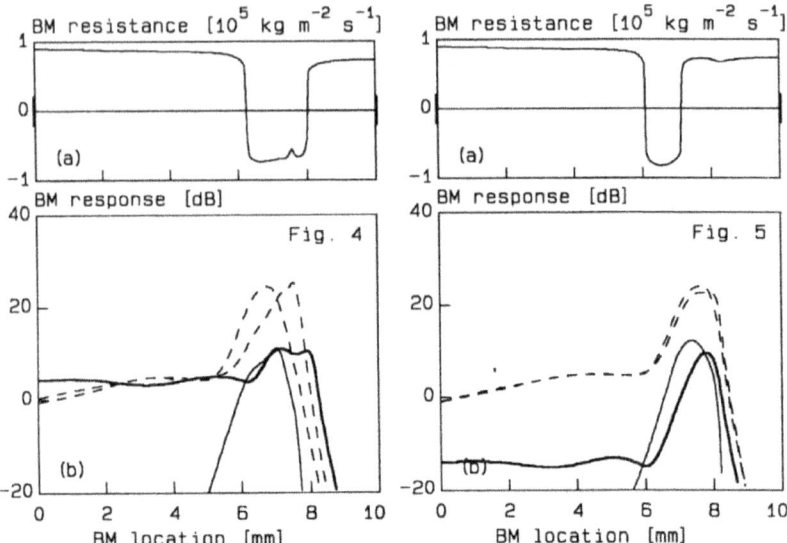

Figure 4. Model responses for two-tone stimulation. Abscissa: Location x along the BM, measured from the stapes. (a) Resistance component of the effective BM impedance for the frequency $2f_1$-f_2 drawn logarithmically for absolute values larger than 10 kg m^{-2} s^{-1}. The small bars on the left and right indicate the region within which the resistance has been plotted in a linear way (with such a slope that continuity is retained). Part (b) shows the velocity responses of the two primaries (thin dashed) and of the DP (thin solid). The thick solid line pertains to the DP pressure response. Primary frequency ratio: 1.077; f_1: 6.5 kHz; f_2: 7 kHz; primary levels both at 50 dB SPL. Parameters are such that our model reproduces the BM velocity response of the guinea pig (Sellick et al., 1982, Fig. 15, curve with closed circles). The reference value of the velocity is 1×10^{-5} m s^{-1}, and for the pressure it is 3.16×10^{-3} N m^{-2}.

Figure 5. Model responses for two-tone stimulation. Same as Fig. 4, but with a primary frequency ratio of 1.02. We see that now the primaries are brought closer together compared to Fig. 4, the pressure response of the DP component decreases at the stapes while the maximal value of the DP sources approximately stays the same. Primary frequency ratio: 1.016; f_1: 6.1 kHz; f_2: 6.2 kHz; primary levels both at 50 dB SPL.

Since in the model the DPs are not filtered after generation, the 'tuning' in Fig. 2 cannot be produced by a second filter. But then, what is the reason for this tuning? One explanation could be reflection at the stapes. This possibility is excluded since the middle ear in the model is reflectionless. The most obvious explanation then is that the 'tuning' is caused by the nonlinearity in the system. To check whether this is true, we compute the

DP emission again as a function of primary frequency ratio, but at lower levels than in Fig. 2. The result is shown in Fig. 3 for which we have used primary levels of 20, 30 and 40 dB SPL. We see that the DP 'tuning' has disappeared completely at the level of 20 dB SPL. In this case the emission increases monotonically as the frequency ratio decreases to 1.0. Because similar stimulus conditions have not yet been exploited in experiments, we have to await future experiments to decide whether 'tuning' exists at low levels. Because there is no need of a second filter to explain the 'tuning' seen at higher levels, this makes it possible (but not strictly necessary) that there will be no 'tuning' at low levels.

At this point we have shown that in a cochlea model without DP filtering the 'tuning' of DPs is similar as that seen in experiments. Secondly, in the model 'tuning' arises at primary levels above 20 dB SPL and is the result of the nonlinearity in the system. Now we would like to know in how far the 'tuning' is caused by suppression of the active mechanism at the DP frequency by the primaries and in how far by the saturation of the DP generation itself.

In Figs. 4 and 5 model responses for two-tone stimulation are shown as a function of distance x from the stapes. Part (a) illustrates the resistance component of the effective CP impedance $Z_{BM}(x;\omega_{DP}) - Z^{(2)}_{OHC}(x;\omega_{DP})$ for the $2f_1$-f_2 DP frequency. Wherever the real part of this impedance is negative, amplification of the DP wave occurs. We refer to the legend for details about the scaling. Part (b) shows the velocity responses of the two primaries (thin dashed lines). The thick solid line pertains to the DP pressure response, and the thin solid line to $P^{(1)}_{OHC}(x;\omega_{DP})$, the DP component in the OHC pressure. In Fig. 4 the primary frequency ratio is 1.077; in Fig. 5 it is 1.016. In both figures the input levels of the two primaries are 50 dB SPL. We see that as primary frequencies are brought closer together, the pressure response of the DP component at the stapes decreases while the magnitude of the DP source distributions approximately stays the same (the location where the maximum of the thin solid line occurs is shifted to the right).

If we compare the BM impedance in Figs. 4(a) and 5(a) we find that between $x = 7$ and 8 mm the BM impedance in Fig. 5 has been compressed so much that it has nearly become passive. The reason of the compression in Fig. 5 is that the primary components are *both* large near the DP resonance place and thus tend to diminish the active process at those locations substantially. In Fig. 4 the responses of the primaries and the resulting suppression are more distributed.

To examine whether the suppression of the active mechanism at the DP frequency might be the cause of the 'tuning' we have artificially removed this compression of the active process at the DP frequency; that is, the 'active' impedance $Z_{OHC}(x;\omega_{DP})$ is now taken as fully active [see Eq. (9) in Kanis and de Boer (1993b)]. (Note that such an action is only possible with our quasilinear method and not with time-domain methods.) The result is shown in Fig. 6. Indeed, the backward travelling wave is amplified more than in Fig. 5 so that the stapes pressure has increased with respect to Fig. 5. (If the same procedure is applied to Fig. 4, there is hardly any difference in emission). However, the increase is not so large that the 'tuning' has fully disappeared as can be seen from Fig. 7 where we have plotted DP emission as a function of primary frequency ratio. The solid line is the 50 dB SPL curve from Fig. 2, and the dashed line is the unsuppressed curve also obtained at 50 dB SPL. The fact that the 'tuning' has not fully disappeared suggests that suppression of the active mechanism at the DP frequency is only partly an explanation of the 'tuning'. Another reason is that the DP generation process saturates at higher levels. However, saturation of the DP generation is not the whole story, since if in a fully-active model a single source of a fixed strength (to be compared with full saturation) is moved from the stapes to its place of resonance, the emission increases (due to amplification of the backward travelling wave): there is no 'tuning'. That the dashed line in Fig. 7 still displays 'tuning' is the result of a complex interplay of wavelets coming from different locations (see thin solid line in Fig. 5 showing the distributed nature of the DP source).

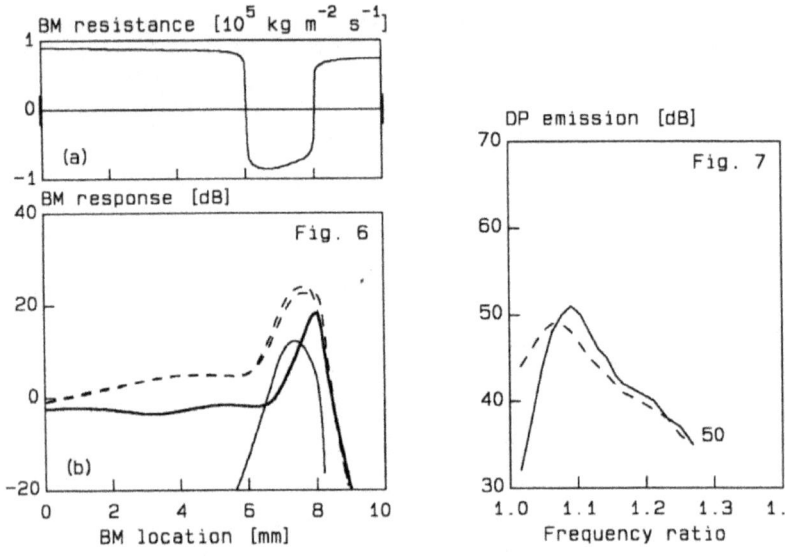

Figure 6. Effect of removal of suppression on DP response pattern. Same as Fig. 5, but now with the suppressive effect of the primaries on the amplification of the DP has artifically been removed. The DP stapes pressure is now larger than in Fig. 5(b) and nearly as large as in Fig. 4(b). The reason is that the backward travelling wave has been amplified more than in Fig. 5(b).

Figure 7. Effect of removal of nonlinearity on DP emissions. The solid curve has been obtained as a function of primary frequency ratio and is the same as the 50 dB SPL curve from Fig. 2. The dashed curve has been obtained with the same parameters but the suppressive effect of the primaries on the amplification of the DP has artifically been removed (see also Fig. 6). We see that the 'tuning' of the dashed curve is much less than for the solid curve.

According to Allen and Fahey (1993) emission curves of DPs of different order all show a peak at the same frequency (if plotted as a function of its own frequency and for f_2 fixed). However, the experiments by Brown and Gaskill (1990a, Fig. 4) and Brown and Williams (1993) show otherwise. Also, in our model the frequency at which these maxima occur tends to increase slightly when n increases, but the effect is not very large. To generate the same DP frequency the primary frequency f_1 must be larger for higher n, so that the suppressive influence of the f_1-primary response on the cochlear amplifier at the DP frequency becomes less. Consequently, for higher n the same amount of suppression in the 'tuning' curve occurs for a slightly higher DP frequency.

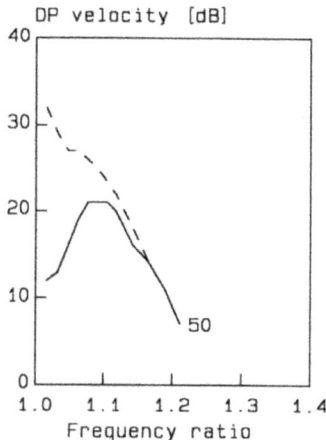

Figure 8. BM velocity of a 6 kHz DP at its characteristic location as a function of primary frequency ratio. The dashed curve has been obtained for the case when suppression of the active mechanism at the DP frequency by the primaries has been removed artificially (as in Fig. 6 and 7); for the solid line suppression has not been removed. Primary levels are at 50 dB SPL.

Instead of viewing the pressure at the eardrum we might also monitor the BM velocity of the DP at its characteristic location (which corresponds with the peak location of a low-level tone at the DP frequency). This can be done by either varying primary frequencies or primary levels. Robles *et al.* (1991) showed that as a function of primary level (and for a primary frequency ratio of 1.1) the DP velocity response first increases but then decreases. The same is seen in our model and the reason is simple: the amplification of the apically travelling wave at the DP frequency becomes less due to suppression by the primaries. The DP velocity response might also be monitored as a function of primary frequency ratio. In Fig. 8 the result (solid line) is shown for model simulations, and we see again first an increase and then a decrease of response. The reason is again: suppression of the active process at the DP frequency by the primary responses. Indeed, the dashed curve which was obtained for the case when suppression by the primaries has been removed artificially (as in Fig. 7) displays a monotonic increase in response for decreasing frequency ratios. The DP velocity response was also monitored as a function of primary frequency ratio by Hall (1974, Fig. 6). Apart from some sharp nulls in the amplitude curve, which were caused by reflection at the stapes, the suppression in his model was not large enough to reduce the DP amplitude at its characteristic place when the primary frequency ratio approached 1.0. The same is true for Zwicker's model (1981, Fig. 15). Buunen and Rhode (1978) and Webert and Mellert (1975) obtained similar results as in Fig. 7 in a neurophysiological and psychoacoustic experiment, respectively.

In experiments (Gaskill and Brown, 1990) it has been found that largest emissions occur when the level of the f_1-primary level is about 10-20 dB higher than the f_2-level. The same has been found in our model. The reason is that more overlap occurs than in the case of equal primary levels.

4.4 Discussion

In this paper we have demonstrated that in order to explain the 'tuning' of DP emissions seen in experiments one does not need a filtering mechanism as hypothesized by some authors (Brown and Gaskill, 1990a; Brown and Williams, 1993; Allen and Fahey, 1993; Neely and Stover, 1993). We have replicated 'tuning' of acoustic DPs in a cochlea model in which OHC-generated pressures are *not* filtered as they are coupled back to the BM. In our model, DP 'tuning' is caused by *saturation* of the active process, and, therefore, our model does not show DP 'tuning' at low levels. Could it be that DP 'tuning' is caused by another mechanism than saturation? In our model that the BM velocity is filtered by the stereocilia before it is used as input to the OHCs. However, irrespective of the place of resonance of this filter, maximal distortion will be generated if both primaries overlap exactly, thus if the primary frequency ratio is 1.0. Therefore, a filter that is 'located' before the OHC input cannot be the cause of DP 'tuning'.

The evidence of DP 'tuning' at low input levels is also small in the cochlea model by Neely and Stover (1993) where DP components in the OHC pressures are filtered before they are coupled back to the BM (Neely, personal communication). However, in their model the DP filtering cannot give rise to DP 'tuning' because the maximum of the filter $F(x;\omega)$ in Eq. (4.A.6) lies at the same location as the peak of the velocity response (at least for low levels). This is not true for the locally active model by Geisler (1991). In that model the filter peak lies at a location that corresponds with the location of the velocity-peak of a one octave higher tone. Thus, in Geisler's model DP 'tuning' at low levels would be possible.

A future experiment performed at low levels might discriminate between a class of models in which DP filtering can be reflected in the acoustic distortion data and a class of models in which that is not the case. It is a fact that the OHCs are embedded in the organ of Corti, and it is therefore likely that the DPs are filtered before they are coupled back on the BM. The question is what the form of this filter is. If a future experiment shows DP tuning at low levels, then the class of models to which Neely and Stover's model belongs is not of the right type.

On the basis of our hypothesis that DP 'tuning' is caused by saturation we may predict that at low levels there is no DP 'tuning'. Experiments have not yet been performed at low enough levels. At this moment we have to await further experiments to give evidence in support of or against the filtering hypothesis. If DP 'tuning' is found in experiments at sufficiently low levels (equal primary levels of 20 dB SPL) this is in favour of Geisler's model (1991) in which a filter is present, otherwise there is no filtering action or the filter cannot be detected with the paradigms used in the experiments.

Finally it is to be noted that the maximum of the 'tuning' curves shifts to higher frequencies for higher primary levels. This nonlinear effect happens in experiments as well as in our simulation study. The effect would not occur if the 'tuning' of DPs were solely caused by linear filtering.

Acknowledgements

This work was supported by the Netherlands Foundation for Scientific Research, project number 810-410-10-1.

Appendix: Filtering of the OHC pressure

In the frequency domain one cochlear section of the models by Neely and Stover (1993), Geisler (1992) and Geisler *et al.* (1993) can be represented by the network shown in Fig. 1 with the active pressure source at the location of the open circle. We will show in this appendix that the pressure over the BM is the sum of the pressure difference in the fluid and the active pressure that is filtered as it is coupled back to the BM. The network is described by three equations:

$$Z_{BM}(x;\omega)v_{BM}(x;\omega) + Z_{RL}(x;\omega)v_{RL}(x;\omega) = -2p(x;\omega), \qquad (4.A.1)$$

$$v_{OC}(x;\omega) = v_{BM}(x;\omega) - v_{RL}(x;\omega), \qquad (4.A.2)$$

and

$$Z_{OC}(x;\omega)v_{OC}(x;\omega) - Z_{RL}(x;\omega)v_{RL}(x;\omega) = P_{OHC}(x;\omega). \qquad (4.A.3)$$

Here all variables and impedances are complex functions of location x and radian frequency ω; $Z_{RL}(x;\omega)$ is the impedance with which the reticular lamina (RL) are attached to the modiolus via the tectorial membrane, $Z_{OC}(x;\omega)$ the impedance of the organ of Corti, and $Z_{BM}(x;\omega)$ the BM impedance. Furthermore, $v_{BM}(x;\omega)$ and $v_{RL}(x;\omega)$ are the BM velocity, and the velocity of the RL, respectively. The difference between these two velocities is $v_{OC}(x;\omega)$, the velocity of amplitude changes of the organ of Corti.

For the distortion products the pressure $P_{OHC}(x;\omega_{DP})$, generated by the OHCs, consists of a nonlinear term $P^{(1)}_{OHC}(x;\omega_{DP})$ that is produced by the primaries, and a term that is produced by the DP itself:

$$P_{OHC}(x;\omega_{DP}) = P^{(1)}_{OHC}(x;\omega_{DP}) + Z_0(x;\omega_{DP})v_{RL}(x;\omega_{DP}). \qquad (4.A.4)$$

with $Z_0(x;\omega_{DP})$ the (quasilinear) transduction impedance of the OHCs at the frequency of the DP. The impedance $Z_0(x;\omega_{DP})$ is influenced by the primaries and by self-suppression. Solving Eq. (4.A.1) for $v_{RL}(x;\omega_{DP})$ leads to

$$|Z_{BM}(x;\omega_{DP}) + Z_{OC}(x;\omega_{DP})F(x;\omega_{DP})|v_{BM}(x;\omega_{DP}) =$$
$$-2p(x;\omega_{DP}) + F(x;\omega_{DP})P_{OHC}^{(1)}(x;\omega_{DP}) \quad , \tag{4.A.5}$$

where $F(x;\omega)$ is given by

$$F(x;\omega) = \frac{Z_{RL}(x;\omega)}{Z_{OC}(x;\omega) + Z_{RL}(x;\omega) + Z_0(x;\omega)} . \tag{4.A.6}$$

On the right-hand side of Eq. (4.A.5), $F(x;\omega)$ is to be interpreted as a filtering of the DPs before they are added to the pressure difference over the BM. In the Neely-Kim model (1986) and our model there is no such filtering of the DPs.

References

Allen, J.B., and Fahey, P.F. (1993). "Evidence for a second cochlear map," in: *Biophysics of Hair Cell Sensory Systems*, edited by H. Duifhuis, J.W. Horst, P. van Dijk, and S.M. van Netten (World Scientific, Singapore), pp. 296-302.

Boer, E. de (1991). "Auditory Physics. Physical principles in hearing theory. III," Phys. Rep. 203, 125-231.

Brown, A.M., and Gaskill, S.A. (1990a). "Can basilar membrane tuning be inferred from distortion measurement?," in *Mechanics and Biophysics of Hearing*, edited by P. Dallos, C.D. Geisler, J.W. Matthews, M.A. Ruggero, and C.R. Steele (Springer-Verlag, Berlin), pp. 164-169.

Brown, A.M., and Gaskill, S.A. (1990b). "Measurement of acoustic distortion reveals underlying similarities between human and rodent mechanical responses," J. Acoust. Soc. Am. 88, 840-849.

Brown, A.M., and Williams, M.W. (1993). "A second filter in the cochlea," in: *Biophysics of Hair Cell Sensory Systems*, edited by H. Duifhuis, J.W. Horst, P. van Dijk, and S.M. van Netten (World Scientific, Singapore), pp. 72-77.

Buunen, T.J.F., and Rhode, W.S. (1978). "Responses of fibers in the cat's auditory nerve to the difference tone," J. Acoust. Soc. Am. 64, 772-781.

Fahey, P.F., and Allen, J.B. (1986). "Characterisation of cubic intermodulation distortion products in the cat external auditory meatus," in: *Peripheral auditory Mechanisms*, edited by J.B. Allen, J.L. Hall, A. Hubbard, S.T.Neely, and A. Tubis (Springer-Verlag, Berlin), pp.314-321.

Gaskill, S.A., and Brown, A.M. (1990). "The behaviour of the acoustic distortion product, $2f_1-f_2$, from the human ear and its relation to auditory sensitivity," J. Acoust. Soc. Am. 88, 821-839.

Geisler, C.D. (1991). "A model for cochlear vibrations based on feedback from motile outer hair cells," Hear. Res. 54, 105-117.

Geisler, C.D., Bendre, A., and Liotopoulos, F.K. (1993). "Time-domain modeling of a nonlinear, active model of the cochlea," in *Biophysics of Hair-cell Systems*, edited by H. Duifhuis, J.W. Horst, P. van Dijk and S.M. van Netten (World Scientific, Singapore), pp. 330-337.

Harris, F.P., Lonsbury-Martin, B.L., Stagner, B.B., Coats, A.C., and Martin, G.K. (1988). "Acoustic distortion products in humans: Systematic changes in amplitude as a function of f_2/f_1 ratio," J. Acoust. Soc. Am. 85, 220-229.

Kanis, L.J., and Boer, E. de (1993a). "The emperor's new clothes: DP emissions in a locally-active nonlinear model of the cochlea," in: *Biophysics of Hair Cell Sensory Systems*, edited by H. Duifhuis, J.W. Horst, P. van Dijk, and S.M. van Netten (World Scientific, Singapore), pp. 304-311.

Kanis, L.J., and Boer, E. de (1993b). "Self-suppression in a locally active nonlinear model of the cochlea: A quasilinear approach," J. Acoust. Soc. Am. 94, 3199-3206.

Kanis, L.J., and Boer, E. de (1994). "Two-tone suppression in a locally active nonlinear model of the cochlea," J. Acoust. Soc. Am. 96, 2156-2165.

Kim, D.O. (1980). "Cochlear mechanics: Implications of electrophysiological and acoustical observations," Hear. Res. 2, 297-317.

Lonsbury-Martin, B.L., Martin, G.K., Probst, R., and Coats, A.C. (1987). "Acoustic distortion products in rabbit ear canal. I. Basic features and physiological vulnerability," Hear. Res. 28, 173-189.

Neely, S.T. (1985). "Mathematical modeling of cochlear mechanics," J. Acoust. Soc. Am. 78, 345-352.

Neely, S.T. and Kim, D.O. (1986). "A model for active elements in cochlear biomechanics," J. Acoust. Soc. Am. 79, 1472-1480.

Neely, S.T. (1993). "A model of cochlear mechanics with outer hair cell motility," J. Acoust. Soc. Am. 94, 137-146.

Neely, S.T., and Stover, L.J. (1993). "Otoacoustic emissions from a nonlinear, active model of cochlear mechanics," in: *Biophysics of Hair Cell Sensory Systems*, edited by H. Duifhuis, J.W. Horst, P. van Dijk, and S.M. van Netten (World Scientific, Singapore), pp. 64-70.

Robles, L., Ruggero, M.A., and Rich, N.C. (1991). "Two-tone distortion in the basilar membrane of the cochlea," Nature 349, 413-414.

Sellick, P.M., Patuzzi, R., and Johnstone, B.M. (1982). "Measurement of basilar membrane motion in the guinea pig using the Mössbauer technique," J. Acoust. Soc. Am. 72, 131-141.

Weber, R., and Mellert, V. (1975). "On the nonmonotonic behavior of cubic distortion products in the human ear," J. Acoust. Soc. Am. 57, 207-214.

Whitehead, M.L., Lonsbury-Martin, B.L., and Martin, G.K. (1992). "Evidence for two discrete sources of $2f_1$-f_2 distortion-product otoacoustic emission in rabbit: I. Differential dependence on stimulus parameters," J. Acoust. Soc. Am. 91, 1587-1607.

Wilson, J.P. (1980). "The combination tone, $2f_1$-f_2, in psychophysics and ear canal recording," in: *Psychophysical, Physiological and Behavioural Studies in Hearing*, edited by G. van den Brink and F.A. Bilsen (Delft Univ. Press, Delft), pp. 43-50.

5

The emperor's new clothes: DP emissions in a locally active nonlinear model of the cochlea[*)]

5.1 Introduction

Simple models of the cochlea generally do not have the tuning properties that they should have in view of basilar membrane (BM) data (e.g., Sellick *et al.*, 1982; Robles *et al.*, 1986). To solve this discrepancy it has been proposed that the cochlea is *locally active* (Kim *et al.*, 1980; de Boer, 1983). Davis (1983) was the first to use, in this context, the name *cochlear amplifier* (CA)[1].

The concept of active wave amplification in the cochlea was met by many with great scepticism, and up to present times the issue does not appear to be settled. On the one hand, several models have been conceived that produce just the right amount of activity at the right place (Neely and Kim, 1983, 1986; Geisler, 1991; Zweig, 1991; Kanis and de Boer, 199-a). On the other hand, researchers who developed alternative (passive) models have had only limited success in matching model responses to modern mechanical data (Kolston, 1988; Kolston and Viergever, 1989; Kolston *et al.*, 1989; Novoselova, 1987, 1989; Allen, 1988, 1991).

The battle between opinions has recently been revived by Allen and Fahey (1992) with an ingenious experiment in which they tried to estimate the power gain in the cochlea from otoacoustic emissions of distortion products. The emission of the cubic difference tone (CDT) with frequency $2f_1$-f_2 was measured as a function of the primary frequencies f_1 and f_2 (f_2>f_1) while keeping both the CDT's frequency and the neural response of a nerve fibre tuned to this frequency constant. It was found that the emissions did not change as primary frequencies were varied. Allen and Fahey concluded that the cochlea

---

[*)] This chapter was presented at the Symposium on *Hair Cell Sensory Systems* in Paterswolde, The Netherlands.

[1] Our definition of the cochlear amplifier is: a collection of physiological devices in the cochlea that amplifies the travelling wave by injecting more energy into a cochlear section than is absorbed in that section. The location of the amplifier depends on the frequency of the BM response; it is generally assumed that the cochlear amplifier lies basally from the site of the response peak. The model considered in this paper is overall stable: power produced in the active region is dissipated elsewhere.

must be passive.

This conclusion is challenged in the present paper. A locally-active nonlinear model of the cochlea is described and the experiment of Allen and Fahey is simulated. Despite a maximum velocity (pressure) gain of more than 40 (20) dB with respect to a passive model, the CDT emissions changed little as primary frequencies were varied over the range used in the experiment. The reason for this result is found and it is concluded that Allen and Fahey's interpretation of their experiment is incorrect.

5.2 Estimating the pressure gain from emissions

The pressure wave with the frequency of the CDT (in this paper we limit ourselved to the CDT with frequency $2f_1$-f_2, $f_2 > f_1$) is mainly generated at the site where both primary responses are large. From this place a backward and forward pressure wave arise. The backward wave travels to the stapes, where it leaves the cochlea and gives rise to an emission that can be measured in the outer ear canal (cf. Kemp, 1979). The forward wave travels to the place of resonance for the CDT where it can be measured neurally (cf. Goldstein and Kiang, 1968) and mechanically (cf. Robles et $al.$, 1991).

Two cases can be distinguished: (**A**) When primary frequencies are far apart ($f_2 / f_1 \approx$ 2), the CDT wave is generated to the left of the CA region and the backward wave will not be influenced by the CA. The forward wave travels through the CA region to the resonance place and is duly amplified. (**B**) When the two primary tones are very close together in frequency ($f_2 / f_1 \approx 1$), the CDT wave originates to the right of the CA region and only the backward wave travels through the territory of the CA. On the basis of comparing case A with case B, Allen and Fahey (1992) have designed their physiological experiment. They measured CDT emissions in the outer ear canal while holding both the frequency and the neural response at the characteristic place of the CDT constant. Although Allen and Fahey expected a difference, in emission of twice the cochlear pressure gain between the two extreme cases A and B, they found essentially no difference. Their conclusion was that the CA is 'as illusory as the emperor's new clothes'. We shall make his clothes visible again.

5.3 Model and method

To simulate the generation of combination tones we use a locally-active nonlinear long-wave model of the cochlea in which outer hair cells (OHCs) form the only source of activity and nonlinearity. The frequency-place distribution of activity is achieved by a mechanism similar to that described by Neely and Kim in their 1986 paper. The resulting cochlear amplifier (CA) lies to the left of the velocity response peak and provides a 40 dB enhancement of the velocity response. The OHCs are assumed to have a compressive action, described by a hyperbolic tangent function. We solve the model in the frequency domain with a $quasi$-$linear$ method (Kanis and de Boer, 1993). With this method distortion products in the BM-velocity response are treated as perturbations[2], and all system variables such as the nonlinear OHC-generated pressure are written as sums of Fourier components. Then, the familiar concept of transfer impedance (defined as the ratio of a pressure to a velocity) is used to solve the cochlear problem in the frequency domain, and this can be done for any component we want. The ultimate solution is obtained by iteration because the transfer impedances depend on the magnitude of the BM velocity.

The method can, in principle, be used for any type of stimulus. In the case of two-tone stimuli, the suppressing interaction between the primary tones is evaluated first. Then, the CDT component in the nonlinear OHC pressure is computed for all locations in the cochlea and the wave equation for the CDT component is solved. The CDT component has negligible influence on the primary components, so we need not evaluate further interactions between the CDT and the primaries. More details are given in Kanis and de Boer (199-b, 199-c).

The cochlea is coupled to the outer ear canal with such an impedance that the model remains stable for all frequencies and that reflections at the stapes are minimized. For this impedance we have taken an approximate form of the local impedance function (Viergever and de Boer, 1987).

5.4 Model results

Figures 1 and 2 show model results for two extreme frequency ratios that were used by Allen and Fahey (henceforth abbreviated A&F). The CDT frequency is chosen as 5 kHz. Fig. 1 illustrates the case where primary frequencies are relatively far apart ($f_2 / f_1 = 1.55$); in Fig. 2 the primary frequencies are closer together ($f_2 / f_1 = 1.09$). Part (a) of both figures shows the resistive component of the BM impedance for both primaries (dashed lines) and the CDT (solid line). Negative excursions indicate local activity (presence of the CA). In part (b) of both figures BM-velocity patterns of the two primary tones (dashed lines) and of the CDT (thin solid line) are shown. The thick solid line is the CDT pressure pattern. The arrow in part (b) indicates where the CDT is generated at most. Note that this is where both primaries are large. In Fig. 1 the arrow points to the left of the CA region for the CDT frequency and in Fig. 2 it points to the middle. The intensities of the primary tones have been chosen such that the CDT velocity peaks are of equal magnitude in both figures, just as in the experiments of A&F.

The figures show that there is not very much difference between the CDT pressures at the stapes in Figs 1 and 2 (thus there will be nearly equal emissions in the ear canal). The same was observed by A&F for these frequency ratios, but they concluded that the cochlea is passive. Our results have been obtained with a model of the cochlea that is *locally active*. This means that the clothes of the emperor are not as illusory as is suggested by A&F. Quite possibly, he is fully dressed.

---

[2] In mechanical experiments with pure tone stimulation it is found that, although the magnitude of the BM-velocity waveform is a nonlinear function of the input level, the waveform itself shows only little distortion (cf. Cooper and Rhode, 1992, Fig. 20).

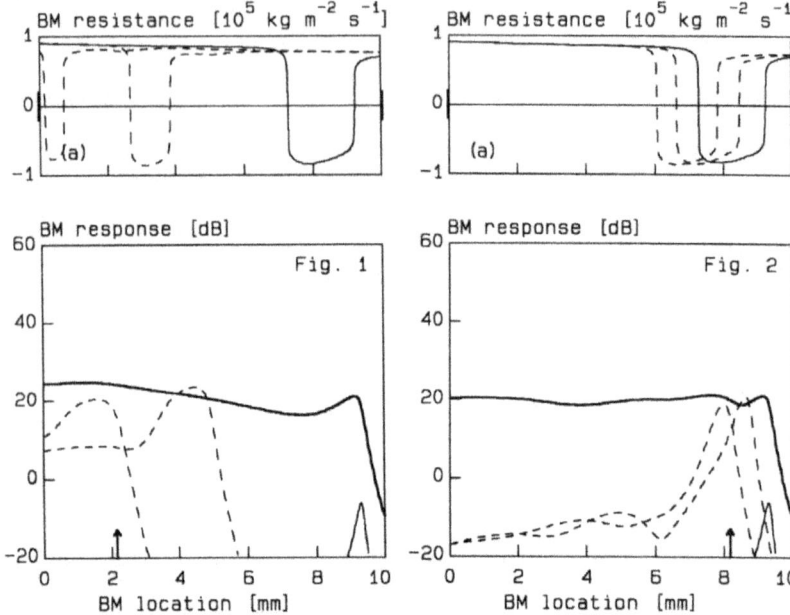

Figures 1 and 2. BM impedance and response in a locally-active nonlinear cochlea model when the input consists of two equally loud tones at level L. In Fig. 1 the primary frequency ratio is 1.55 and primary levels are 53 dB SPL; for Fig. 2 the ratio is 1.09 and primary levels are 34 dB SPL. (a) Resistive component of the BM impedance for both primaries (dashed lines) and the CDT (solid line). Large excursions are shown compressed in this figure. (b) BM velocity patterns of the two primary tones (dashed lines) and of the CDT (thin solid line). The thick solid line is the CDT pressure pattern. The arrow indicates where the CDT is generated at most. The reference value of the velocity is 1×10^{-5} m s^{-1}, and for the pressure it is 3.16×10^{-4} N m^{-2}. Note that the velocity response at the peak is the same in the two figures.

5.5 Analysis and method

Allen and Fahey (A&F) expected a difference in CDT emission of twice the cochlear gain (see our brief description in section 2), but they measured only a small difference. Why would that be so? The first thing that strikes the eye is that Fig. 2 does not represent case B described in section 2. Bearing in mind that Figs 1 and 2 were generated with approximately the same frequency ratios as used by A&F, it is clear that *A&F did not really compare case A with case B*. Due to experimental limitations they were not able to make the primary frequency ratio much smaller than 1.09. Whether or not the cochlea is active, cannot be deduced from their experimental results.

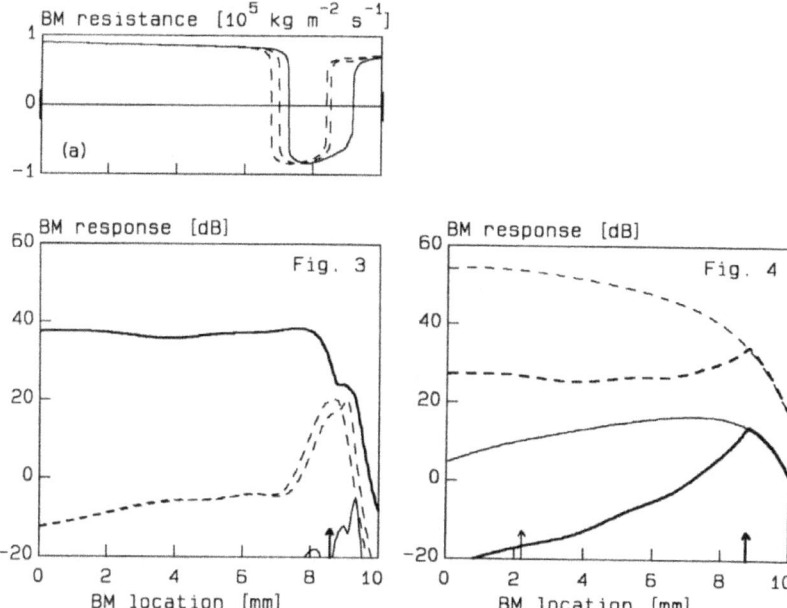

Figure 3. Same as Figs 1 and 2, but with a primary frequency ratio of 1.04 and primary levels of 40 dB SPL.

Figure 4. BM responses in a passive cochlea model which is excited internally by one sinusoidal pressure source with a frequency of 5 kHz. Solid lines indicate velocity responses and dashed lines pressure responses. For the thin lines the source lies near the stapes (thin arrow), and for the thick lines the source lies at the location of the thick arrow. The pressure has been attenuated by 60 dB with respect to the velocity response.

In a model the range of experimental variables can easily be extended. In Fig. 3 results are shown when the model is solved for a frequency ratio of 1.04. The arrow now points to the right of the CA region; this confirms that we are really dealing with case B. When we compare Fig. 3 with Fig. 1, we find a difference in emission of about 10 dB. This value (which is a *prediction* that can be tested experimentally) is certainly *not as large* as A&F expected.

In order to understand why the difference is so small we analyze the course of the pressure wave in a *passive* (and linear) cochlear model. Figure 4 shows two pressure (dashed line) and velocity (solid line) responses in a passive model which is excited internally by a single sinusoidal pressure source at two different locations. The thin lines correspond to responses for which the source lies at the location of the thin arrow and for the thick lines the pressure source lies at the thick arrow. To get maximal correspondence with the nonlinear model results, the thin and thick arrows lie at the same locations as the arrows in Figs 1 and 3, respectively. In both cases the strength of the source has been chosen so that the velocity is the same at the peak location for the CDT response in Figs 1-3.

If there would be no *power dissipation* of the travelling wave, the two pressures at the stapes location would be equal. However, dissipation does occur in the region between the thin and thick arrow, and this amounts to a substantial difference (minus 26 dB in our model, see Fig. 4) between the two pressures. In a *locally-active* model this dissipative effect is also present, since the waves traverse a region where the model is not active and where dissipation does play a role. In our model the CDT emission is greatly reduced by this effect.

When we would have placed the two arrows in Fig. 4 at the locations of the arrows in Figs 1 and 2, the difference in emission in our model would be minus 18 dB. According to this result, A&F would have measured a considerable decrease in CDT emission in a passive (but nonlinear) cochlea when the primary frequency ratio were decreased from 1.55 to 1.09. As this was not the case one might conclude from their experiment that the cochlea is active.

Finally, it should be noted that the relation between the difference in CDT emissions and the value of the CA gain is not straightforward. Several reasons can be given. First, the CDT emissions are influenced by dissipation. Secondly, the influence of the CA on the response depends on the travel direction of the cochlear wave, since the phase of the OHC-generated pressure is related with the phase of the cochlear wave. Furthermore, the CA of the CDT is suppressed by the primaries when they are strong enough (see Fig. 3b), and the CDT is generated by more than one pressure source in Figs 1-3. Thus, it will be difficult to estimate the exact value of the cochlear gain by varying the frequency ratio of the primaries.

5.6 Conclusions

By replicating the experiments of Allen and Fahey (1992) with a computer model, we arrived at the same results as presented in their paper. Only, our results have been obtained with a locally-active model of the cochlea, while Allen and Fahey surmised that similar results could only be obtained for a passive cochlea. When we decreased in our model the ratio of primary frequencies, first the emission of the CDT component decreased due to dissipation, but then increased by the action of the cochlear amplifier. Due to dissipation this increase remained limited. For instance, when the frequency ratio in our model was changed from 1.55 to 1.04, the difference in emission was only 10 dB. However, it will be difficult to estimate the exact value of the cochlear gain in this way, since the relation between the emissions and the value of the CA gain is not straightforward. At the present time, the frequency response of the mechanical data can only be compatible with cochlear models that have a fair amount of wave amplification. And that colours the emperor's clothes clearly red.

Acknowledgements

This work was supported by the Netherlands Organization for Pure Research.

References

Allen, J.B. (1988). "Cochlear signal processing ," in: *Physiology of the ear*, edited by A.F. Jahn and J. Santos-Sacchi (Raven Press, New York), pp. 243-270.

Allen, J.B. (1991)."Modeling the noise damaged cochlea," in: *Mechanics and Biophysics of Hearing*, edited by P. Dallos, C.D. Geisler, J.W. Matthews, M. Ruggero and C.R. Steele (Springer, Berlin), pp. 362-269.

Allen, J.B. and Fahey, P.F. (1992). "Using acoustic distortion products to measure the cochlear amplifier gain on the basilar membrane," J. Acoust. Soc. Am. 96, 178-188.

Boer, E. de (1983). "No sharpening? A challenge for cochlear mechanics," J. Acoust. Soc. Am. 73, 567-573.

Cooper, N.P. and Rhode, W.S. (1992). "Basilar membrane mechanics in the hook region of cat and guinea-pig cochleae: Sharp tuning and nonlinearity in the absence of baseline position shifts," Hear. Res. 63, 163-190.

Davis, H. (1983). "An active process in cochlear mechanics," Hear. Res. 9, 79-90.

Geisler, C.D. (1991). "A cochlear model using feedback from motile outer hair cells," Hear. Res. 54, 105-117.

Goldstein, J.L. and Kiang, N.Y.S. (1968). "Neural correlates of the aural combination tone 2f1-f2," Proc. IEEE 56, 981-992.

Kanis, L.J., and Boer, E. de (1993). "Self-suppression in a locally active nonlinear model of the cochlea: A quasilinear approach," J. Acoust. Soc. Am. 94, 3199-3206.

Kanis, L.J., and Boer, E. de (1994a). "Two-tone suppression in a locally active nonlinear model of the cochlea," J. Acoust. Soc. Am. 96, 2156-2165.

Kanis, L.J., and Boer, E. de (1994b). "Frequency dependence of acoustic distortion products in a locally active model of the cochlea," submitted to the Journal of the Acoustical Society of America.

Kemp, D.T. (1979). "Evidence of mechanical nonlinearity and frequency selective wave amplification in the cochlea," Arch. Otorhinolaryngol. 224, 37-45.

Kim, D.O., Neely, S.T. , Molnar, C.E. and Matthews, J.W. (1980). "An active cochlear model with negative damping in the partition: Comparison with Rhode's ante- and post-mortem observations," in: *Psychophysical, Physiological and Behavioural Studies in Hearing*, edited by G. van den Brink and F.A. Bilsen (Delft Univ. Press, Delft), pp. 7-14.

Kolston, P.J. (1988). "Sharp mechanical tuning in a cochlear model without negative damping," J. Acoust. Soc. Am. 83, 1481-1487.

Kolston, P.J. and Viergever, M.A. (1989). "Realistic basilar membrane tuning does not require active elements," in: *Cochlear Mechanisms, Structure, Function and Models*, edited by J.P. Wilson and D.T. Kemp (Plenum Press, London), pp. 415-424.

Kolston, P.J., Viergever, M.A., de Boer, E. and Diependaal, R.J. (1989). "Realistic mechanical tuning in a micromechanical cochlear model," J. Acoust. Soc. Am. 86, 133-140.

Neely, S.T. and Kim, D.O. (1983). "An active cochlear model showing sharp tuning and high sensitivity," Hear. Res. 9, 123-130.

Neely, S.T. and Kim, D.O. (1986). "A model for active elements in cochlear biomechanics," J. Acoust. Soc. Am. 79, 1472-1480.

Novoselova, S.M. (1987). "The influence of mass and stiffness cross-sectional distribution on basilar membrane tuning," Sov. Phys. Acoust. 33, 363-365.

Novoselova, S.M. (1989). "A possibility of sharp tuning in a linear transversally inhomogeneous cochlear model," Hear. Res. 41, 125-136.

Robles, L., Ruggero, M.A. and Rich, N.C. (1986). "Basilar membrane mechanics at the base of the chinchilla cochlea. I. Input-output functions, tuning curves, and response phases," J. Acoust. Soc. Am. 80, 1364-1374.

Robles, L., Ruggero, M.A. and Rich, N.C. (1991). "Two-tone distortion in the basilar membrane of the cochlea," Nature 349, 413-414.

Sellick, P.M., Patuzzi, R. and Johnstone, B.M. (1982). "Measurement of basilar membrane motion in the guinea pig using the Mössbauer technique," J. Acoust. Soc. Am. 72, 131-141.

Viergever, M.A. and de Boer, E. (1987). "Matching impedance of a nonuniform transmission line: Application to cochlear modeling," J. Acoust. Soc. Am. 81, 184-186.

Zweig, G. (1991). "The impedance of the Organ of Corti," in: *Mechanics and Biophysics of Hearing*, edited by P. Dallos, C.D. Geisler, J.W. Matthews, M. Ruggero and C.R. Steele (Springer, Berlin), pp. 362-269.

Comment by Allen   In the introduction it says: 'Simple models of the cochlea generally do not have the tuning properties that they should have in view of basilar membrane (BM) data (e.g., Sellick *et al.*, 1982; Robles *et al.*, 1986).'

In this comment we would like to compare the responses published in Sellick *et al.* (1982a), Sellick *et al.* (1982b) and the more recent Robles *et al.* (1986) data given in

Ruggero *et al.* (1990, Fig. 18).
- Sellick *et al.* (1982a) versus Sellick *et al.* (1982b). If one were to plot the 1982a displacement against the 1982b displacement there is more than 25 dB difference that is systematically increasing for decreasing frequencies; at CF there is a 15-20 dB difference. This difference is too large to attribute to experimental error. The question is, why are they so different?

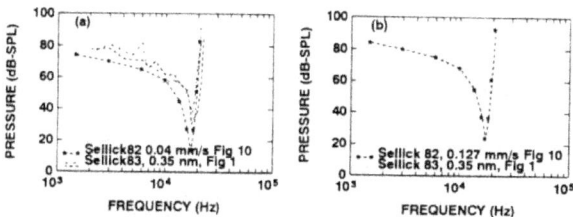

Figure 1: Comparison between Sellick *et al.* (1982a) velocity and Sellick *et al.* (1982b) displacement.

In Fig. 1 we see a somewhat different comparison of Sellick *et al.* (1982a) with Sellick *et al.* (1982b). The left panel compares *iso-velocity* data in 1982a from Fig. 10 to *iso-displacement* data from 1982b, Fig. 1. The group of three curves corrspond to the small source on the edge of the BM. This was the condition they concluded had the smallest effect on the response curves. Large and small sources placed in the middle of the BM, and large sources placed on the edge of the BM gave different responses. The right panel shows the same data except the 1982a data has been shifted by 10 dB upward, corresponding to a less sensitive condition, to give the maximum overlap of the responses. Only one of the three 1982b curves is shown in this panel.

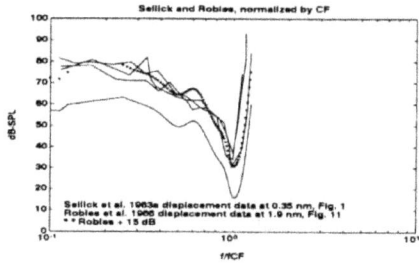

Figure 2: Comparison between Sellick *et al.* (1982b) and Ruggero *et al.* (1990).

It is clear that the 1982a and 1982b data differ by a large amount, namely 6 dB/octave plus a 10 dB shift. Which data are we to believe, 1982a or 1982b? By 1982b they had discovered that the source could influence the measurement, and they had presumably refined the measurement technique.
- Sellick *et al.* (1982b) versus Ruggero *et al.* (1990). In Fig. 2 we compare Sellick *et al.* (1982b) iso-displacement data at 0.35 nm, Fig. 1 to that derived from Ruggero *et al.* (1990) at 1.9 nm. The stars are the Ruggero data shifted up by 15 dB. At first glance this appears like a nice fit as the curves have a similar frequency response. A 15 dB shift would correspond to about 10.7 nm assuming linearity. When the two curves are compared on an absolute scale however, they are 30 dB apart.

The source used by Ruggero *et al.* (1990) was $80\times80\,\mu\text{m}^2$, while the sources used by Sellick *et al.* were $50\times50\,\mu\text{m}^2$ in 1982a and $20\times50\,\mu\text{m}^2$ and $60\times86\,\mu\text{m}^2$ in 1982b. Sellick *et al.* found that the larger sources influenced their measurements.

In conclusion, which data should we model? None of the data in the literature seems to stand the test of comparison. The response in the 'tail' region where we expect a linear response seems equally uncertain.

References

Robles, L., Ruggero, M.A. and Rich, N.C. (1986). "Basilar membrane mechanics at the base of the chinchilla cochlea. I. Input-output functions, tuning curves, and response phases," J. Acoust. Soc. Am. 80, 1364-1374.

Sellick, P.M., Patuzzi, R. and Johnstone, B.M. (1982a). "Measurement of basilar membrane motion in the guinea pig using the Mössbauer technique," J. Acoust. Soc. Am. 72, 131-140.

Sellick, P.M., Patuzzi, R. and Johnstone, B.M. (1982b). "Comparison between the tuning properties of inner hair cells and basilar membrane motion," Hear. Res. 10, 93-100.

Ruggero, M.A., Robles, L., and Rich, N. (1990). "Middle-ear responses in the chinchilla and its relationship to mechanics at the base of the cochlea," J. Acoust. Soc. Am. 87, 1612-1629.

Reply It is necessary to define what we mean by 'tuning'. Considering the cochlea as linear, turn the measured iso-velocity and iso-displacement curves upside down and inspect the resulting frequency-response curves. These curves have a general trend at low frequencies: depending on the way they have been measured, the response rises with zero to 6 dB per octave. Imagine this trend to be continued to higher frequencies, and observe then that the response has a high and broad peak above the low-frequency course. In the left part of Fig. 1 the height of the peak (dashed lines) is 40 dB, for the right-hand part it is 43 dB. For the Robles *et al.* data, Fig. 2, it is smaller, 29 dB.

Peaks of this size *and* shape cannot be obtained with what we have called a 'simple model of the cochlea' [that type of model is more extensively described in section 1 of de

Boer's paper (1993)]. The meaning and validity of the above statement has been the subject of many discussions over the last ten years. Arguments pro and con have been reviewed (albeit not exhaustively) in de Boer's paper (1993). The present paper deals with one aspect of that discussion.

References

Boer, E. de (1993). "Some like it active," in: *Biophysics of Hair Cell Sensory Systems*, edited by H. Duifhuis, J.W. Horst, P. van Dijk, and S.M. van Netten (World Scientific, Singapore), pp. 3-22.

Comment by Allen and Fahey For the record, Allen and Fahey (A&F) do not allude to the "emperor's clothes". This reference is to a New York times article, written by Malcolm Brown in the Science Times of June 9, 1992, page C1, where Brown quotes Allen as referring to the cochlear amplifier (CA) to being similar to the emperor's clothes. Allen was more than a little surprised of this quote when he first read the article because he did not remember referring to any emperors.

On a more serious note: the A&F experiment found no evidence for distributed basilar membrane amplification, as stated in the abstract of A&F, to within '2 mm of the place corresponding to the frequency being measured'. This is consistent with Figs. 1, 2 and 3 of Kanis and de Boer (K&DB) where the region of gain is localized to 1.7 mm basal to the characteristic place. Indeed, these figures support the A&F measurement strategy and K&DB show in their Fig. 3 a gain sensitivity when the DP source is well within the amplification region. Both A&F and K&DB seem to agree that, if there is amplification, it is local to the characteristic place. The question then becomes: Can local gain model the data? This question is addressed in Figs. 9 and 10 (of A&F) which argue where the CA must be relative to the CF. These arguments are based on the slope change in the neural excitation patterns. We specifically studied where the DP source was relative to this region, and showed that the source was inside the estimated CA region for the smallest f_2/f_1 ratio. K&DB have placed their source well inside this region, and do not see a difference in pressure. How can this be?

The results of K&DB only consider two general locations for the source of DP energy, near and far from the place. They also use a model with 13 dB of power loss. It appears that their results are a special case where the CA gain is cancelled by this 13 dB propagation loss. In other words, the pressures shown in Figs. 1 and 2 seem to result from the cancellation of two big numbers, the CA gain and the propagation losses. If this is the case, then this represents a significant difference from the A&F results, where the source was varied continuously from high frequencies to low. If K&DB were to move the source continuously through the region basal to the CF, we could better see the whole picture, as in the A&F case. A&F results, experimental measurement, and model calculations, lead us to believe that cochlear losses are small.

But there is a much more fundamental question to be answered before modelling, with or without cochlear amplifiers. What are we modelling, the neural responses or the mechanical response of the cochlear partition? There seems to have arisen the belief that these are the same. We believe that the experimental evidence does not support the common assumption that the neural and mechanical tuning are the same.

Reply It was certainly not our intention to raise the suggestion that the reference to the emperor's new clothes was taken from A&F's paper: otherwise we would have put quotes around it.

We modelled a cochlea showing a pronounced and broad response peak (see reply to comment by Allen), and could do this with an 'active' region, 1.7 mm wide. When f_2/f_1 equals 1.09, the effective source of the DP is inside the active region, see Fig. 2 of our paper. It is because the major part of wave amplification occurs in the right-hand part of this region that the DP emission pressures in Figs. 1 and 2 turn out to be nearly equal.

The loss of 26 dB occurs, as we state, in a passive model. In the active region of an active model the loss would be more than compensated by the wave-amplification system. The left-going pressure wave in Fig. 4 (thick dashed line) shows, outside the active region, approximately the same course as the thick line in Fig. 3. Over this region the loss is not zero but it remains small. It forms one of the reasons why the 'gain' in DP emission does not become as large as two times the 'cochlear gain' (which actually should be: the pressure gain, not the power gain).

In the case under discussion: if the cochlea were passive, Allen and Fahey would have measured a DP pressure that is smaller in the case $f_2/f_1 = 1.09$ than when f_2/f_1 equals 1.55. It is difficult to design a passive long-wave model with a substantially smaller loss than the 26 dB that we have in the passive version of our model. Again, since Allen and Fahey did not observe such a reduction, we maintain that the cochlea is active.

As to your last question, we have assumed a linear relation between the BM-velocity response and the neural response; in other words, equal velocity responses at the DP peak give rise to equal neural responses.

Comment by Goldstein: Does your nonlinear model simulate the known data on differences in growth of two-tone suppression for low-side and high-side suppressor tones?

Reply Yes, it does. The reason is the following: the influence of the cochlear amplifier (CA) on the BM response is largest for locations near BM resonance. In other words, suppression of the CA has less effect on the probe response when the CA suppression occurs for a location far from the probe resonance compared to a location near probe resonance. This also holds for *changes* in reductions of the activity (whcich occurs when the suppressor level is changed). Since a high-side suppressor influences the CA (belonging to the frequency of the probe) basal to the probe peak, while for low-side suppressors the influence is largest near the probe peak, the reason for the suppression-slope difference you refer to becomes obvious.

6

Comparing frequency-domain with time-domain solutions for a locally-active nonlinear model of the cochlea

Abstract   In previous papers (Kanis and de Boer, 1993a, 1993b, 1994a, 1994b) we have described and applied an approximation method, called the quasilinear method, to solve stable nonlinear locally-active cochlea models in the frequency domain. The quasilinear method consists of splitting the cochlear waveforms into primary components and higher-order components that are treated as perturbations. In order to check the accuracy of the quasilinear solution method we made a comparison between frequency-domain and time-domain solutions of a cochlea model in which all activity and nonlinearity resides in the outer hair cells. We used the time-domain method described by Diependaal *et al.* (1987). We did the comparison for single-tone as well as for two-tone stimulation. In most cases the match between the two solutions is excellent. Only in some cases do the responses computed with the two methods differ, but this does not affect the outcome of the analyses performed with the quasilinear method.

6.1 Introduction

In previous papers (Kanis and de Boer, 1993a, 1993b, 1994a, 1994b) we have described and used an approximation method to solve stable nonlinear locally-active cochlea models in the frequency domain. One advantage of this, quasilinear, method over existing time-domain solution methods is a substantial gain in computation time. Another, more important, advantage is that we can use concepts from linear system theory such as impedance and phase angle. This allows us to acquire insight into the mechanism of nonlinearity: several phenomena that were not yet fully understood have been clarified now. Furthermore, we can perform *Gedanken* experiments that are not realizable in the time domain. For instance, in a previous paper (Kanis and de Boer, 1994) we artificially cancelled the suppression of the active mechanism of the $2f_1$-f_2 combination tone by the primaries in order to estimate the effect of the suppression on the 'tuning' of acoustic distortion products.

The quasilinear method is based upon the fact that in a stable model, for stimulation with a periodic stimulus, the basilar membrane oscillates periodically with the same frequency.[1] Thus, the basilar membrane response can be perfectly analysed at every location in terms of its Fourier components. The same is true for the OHC-generated pressure. In the quasilinear method only the fundamental components are carried through the computations, the other ones are treated as perturbations. Because the Fourier components of the OHC-generated pressure are nonlinear functions of the basilar membrane response the model is solved by iteration. In each iteration step, we solve a linear model for the relevant Fourier component.

In an earlier report we showed that for single-tone stimulation the higher-order distortion components do not noticeably influence the primary response (Kanis and de Boer, 1993b, Appendix B). This means that we do not need to consider higher-order products in order to compute the primaries, so that a limited amount of iteration steps suffices. We assumed the same to be true in our paper about two-tone stimulation (Kanis and de Boer, 1994). In the present paper we would like to check whether these assumptions are true by comparing time-domain with frequency-domain solutions. We will check whether we can neglect the presence of the higher-order products in computing the primaries. We will do the comparison for single- as well as for two-tone stimulation. Furthermore, we will check whether the time-domain and the quasilinear method give the same results for combination tones (see Kanis and de Boer, 1993a, 1994b). Another reason to perform the comparison is that the success of the quasilinear method raises the question whether it would be useful in other cases too, when complex stimuli, such as noise or impulse signals, are taken as stimuli. Before we dare to answer this question we must have more quantitative data on the accuracy of the quasilinear method.

We will see that the deviations between the two model solutions are small enough to be certain about the validity of the conclusions made in previous papers (Kanis and de Boer, 1993a, 1993b, 1994a, 1994b) and of possible conclusions to be made in future papers that will discuss model behaviour for more complex signals.

6.2 Model and method

We have modelled the cochlea as a straight fluid-filled narrow tube divided into two rectangular scalae by a movable partition called the basilar membrane (BM). It is assumed that all longitudinal coupling is through the fluid, and that only long waves exist. Our model is based on the model described by Neely and Kim (1986); a 'secondary resonator' is used to ensure the proper place-frequency distribution of OHC activity. The time-domain implementation follows the method outlined by Diependaal et al. (1987). The partial differential equation that describes the BM response is

$$p_{xx}(x,t) + \rho a_{BM}(x,t) / h = 0, \tag{6.1}$$

---

[1] In an unstable cochlea model, such as the panergic model by Duifhuis et al. (1986) and Diependaal and Viergever (1988), spontaneous oscillations with aperiodic limit cycles may arise which renders their model unsusceptible to our method.

where $p(x,t)$, a function of location x and time t, is the dynamic pressure in the fluid, $a_{BM}(x,t)$ the acceleration of the BM, ρ the fluid density ($\rho=10^3$ [kg m^{-3}]) and h the height of the scalae ($h=10^{-3}$ [m]). The subscript x indicates the derivative with respect to location. Equilibrium over the partition is described by

$$-2p(x,t) + P_{OHC}(x,t) =$$
$$m_{BM}(x)a_{BM}(x,t) + r_{BM}(x)v_{BM}(x,t) + k_{BM}(x)u_{BM}(x,t), \tag{6.2}$$

where $m_{BM}(x)$, $r_{BM}(x)$, $k_{BM}(x)$ are the mass, resistance and stiffness of the BM, and $v_{BM}(x,t)$ and $u_{BM}(x,t)$ the velocity and displacement of the BM. The pressure difference over the membrane consists of the fluid pressure difference, $-2p(x)$, and a pressure, $P_{OHC}(x,t)$, generated by the outer hair cells (OHC). Movements of the BM lead to relative movements between the upper side of the OHCs and the tectorial membrane (TM) thereby displacing the stereocilia of the outer hair cells:

$$v_{SC}(x,t) = v_{BM}(x,t) - v_{TM}(x,t), \tag{6.3}$$

where $v_{SC}(x,t)$ and $v_{TM}(x,t)$ denote the velocities of the stereocilia and TM, respectively. The lever gain that exists between (transversal) movements of the BM and (radial) movements of the cuticular plate (at the upper side of the OHCs) was taken equal to unity. If we let $a_{TM}(x,t)$ represent the acceleration of the TM, and $u_{SC}(x,t)$ the displacement of the stereocilia, the second equation of motion can be written as

$$m_{TM}(x)a_{TM}(x,t) = r_{SC}(x)v_{SC}(x,t) + k_{SC}(x)u_{SC}(x,t), \tag{6.4}$$

where the resistance and stiffness of the stereocilia and the mass of the TM are denoted by $r_{SC}(x)$, $k_{SC}(x)$, and $m_{TM}(x)$, respectively. The OHC pressure is finally computed from

$$P_{OHC}(x,t) = r_0(x)v_{SC}(x,t) + k_0(x)u_{SC}(x,t), \tag{6.5}$$

with $k_0(x)$ a parameter with the dimension of stiffness, and $r_0(x)$ a parameter with the dimension of resistance.

The parameters have been given the following values:

$$m_{BM}(x) = m_{BM}, \tag{6.6}$$

$$r_{BM}(x) = \delta m_{BM}\omega_{loc}(x), \tag{6.7}$$

$$k_{BM}(x) = m_{BM}(x)\omega_{loc}^2(x), \tag{6.8}$$

$$r_{SC}(x) = \eta r_{BM}(x), \tag{6.9}$$

$$k_{SC}(x) = \sigma^2 k_{BM}(x),\tag{6.10}$$

$$m_{TM}(x) = m_{BM}(x),\tag{6.11}$$

$$r_0(x) = c_{00}\varepsilon r_{BM}(x),\tag{6.12}$$

$$k_0(x) = c_{00}k_{BM}(x)\tag{6.13}$$

with

$$\omega_{loc}(x) = \sqrt{\frac{k_{BM}(0)}{m_{BM}(0)}}\, \exp(-\alpha x / 2),\tag{6.14}$$

and $m_{BM} = 0.5$ [kg m^{-2}], the damping coefficient $\delta = 0.4$, $\eta = 0.35$, $k_{BM}(0) = 10^{10}$ [kg m^{-2} s^{-2}], $\sigma^2 = 0.49$, $\varepsilon = 2.5$, $c_{00} = 0.11$, and $\alpha = 3 \times 10^2$ [m^{-1}]. The coefficent σ indicates how much the secondary resonance has been shifted with respect to the BM resonance. In our model the secondary resonance lies about half an octave below the BM resonance. Note that, if one finds the values for $r_{SC}(x)$, $k_{SC}(x)$ and $m_{TM}(x)$ too large, both sides of Eq. (6.4) can be multiplied with a constant smaller than unity without changing any of the following equations. This is possible because in Eq. (6.2) there is no term that describes the parallel circuit formed by the mechanics of the TM and the OHC stereocilia.

In what follows we will shortly describe the time-domain method developed by Diependaal et al. (1987).

We would like to write Eq. (6.1) as an equation that has on its left-hand side terms with pressure $p(x)$ and on the right-hand side known variables such as the velocity and the displacement corresponding to the two degrees of freedom. Then, substitution of

$$g(x,t) = r_{BM}(x)v_{BM}(x,t) + k_{BM}(x)u_{BM}(x,t) - P_{OHC}(x,t)\tag{6.15}$$

into Eq. (6.1) leads to

$$p_{xx}(x,t) - (2\rho / hm)p(x,t) = (2\rho / hm)g(x,t).\tag{6.16}$$

This equation can be written as a finite-difference matrix equation that has to be solved for $p(x,t)$ at each time instant, given the function $g(x,t)$. The accelerations $a_{BM}(x,t)$ and $a_{TM}(x,t)$ are computed from

$$a_{BM}(x,t) = (-p(x,t) - g(x,t)) / m,\tag{6.17}$$

and

$$a_{TM}(x,t) = \frac{\eta_{BM}(x)(v_{BM}(x,t) - v_{TM}(x,t)) + \sigma^2 k_{BM}(x)(u_{BM}(x,t) - u_{TM}(x,t))}{m_{BM}(x)},$$

$$(6.18)$$

respectively. Then, they are, together with $v_{BM}(x,t)$ and $v_{TM}(x,t)$, integrated with the fourth-order Runge-Kutta method. After each integration step, $g(x,t)$ is updated with Eq. (6.15), substituted in Eq. (6.16), and so on.

When steady state has been reached we record the velocity waveform at 90 different locations along the cochlear partition during one or more milliseconds (depending on the primary frequencies). In order to analyse the waveforms in terms of their Fourier components, we need an integer number of cycles in one time frame of the Fast Fourier Transform. We record 1 millisecond of the waveform if the stapes is stimulated by two tones with a frequency of, for instance, 6 and 7 kHz. In the case of stimulation by a 6.25 and 6.5 kHz tone we need to record 4 milliseconds in order to get an integer number of cycles in one time frame. The timestep is taken equal to $1 \times 10^{-3}/N$ [s]. In our computations it was sufficient to take N equal to 128.

The boundary conditions are similar as those described in Diependaal *et al.* (1987) but with the middle ear given by

$$T_m p_0(t) - p(0,t) = -R_m v_{stapes}(t) - M_m a_{stapes}(t),$$

$$(6.19)$$

where the lever gain T_m has the (large) value of 355 to maintain consistency with the data of Sellick *et al.* (1982, Fig. 15, curve with closed circles). The input pressure is indicated by $p_0(t)$, $R_m = 63 \times 10^3$ [kg m^{-2} s^{-1}], $M_m = 0.197$ [kg m^{-2}], and $v_{stapes}(t)$ and $a_{stapes}(t)$ are the velocity and the acceleration of the stapes, respectively. The values of R_m and M_m have been chosen so that they are comparable in magnitude with the resistance and reactance parts of the middle ear impedance used for the frequency-domain implementation in Kanis and de Boer (1993b). The other equation describing the movement of the stapes is

$$p_x(0,t) = -\rho a_{stapes}(t).$$

$$(6.20)$$

Implementation of the long-wave model in the frequency domain was described in Kanis and de Boer (1993b). The boundary conditions are described by the frequency-domain equivalents of Eqs. (6.19) and (6.20). In the following we will show that the time-domain equations (6.3)-(6.5) lead to the OHC impedance in Eq. (9) of the cited paper. In the frequency domain all variables and impedances are complex functions of location x and radian frequency ω. The complex variable $v_{BM}(x;\omega)$ (which will be called the *primary BM-velocity response*) is the first-order Fourier component of $v_{BM}(x,t)$ and defined by

$$v_{BM}(x;\omega) = -2i \int_0^T \frac{dt}{T} v_{BM}(x,t) \exp(i\omega t),$$

$$(6.21)$$

where T is equal to $1/f_0$, where f_0 is the frequency of repetition of the BM velocity. Similar expressions hold for the primary component of the pressure $p(x,t)$ and $P_{OHC}(x,t)$ and all other variables. If the first-order Fourier transforms of the time-domain equations are taken, Eqs. (6.2)-(6.5) are replaced by the following equations:

$$-2p(x;\omega) + P_{OHC}(x;\omega) = Z_{BM}(x;\omega)v_{BM}(x;\omega), \tag{6.22}$$

$$Z_{TM}(x;\omega)v_{TM}(x;\omega) = Z_{SC}(x;\omega)v_{SC}(x;\omega), \tag{6.23}$$

$$v_{SC}(x;\omega) = v_{BM}(x;\omega) - v_{TM}(x;\omega), \tag{6.24}$$

$$P_{OHC}(x;\omega) = Z_0(x;\omega)v_{SC}(x;\omega) \tag{6.25}$$

with the complex impedances given by

$$Z_{BM}(x;\omega) = i\omega m_{BM}(x) + r_{BM}(x) + k_{BM}(x) / (i\omega), \tag{6.26}$$

$$Z_{TM}(x;\omega) = i\omega m_{TM}(x;\omega) \tag{6.27}$$

$$Z_{SC}(x;\omega) = r_{SC}(x) + k_{SC}(x) / (i\omega). \tag{6.28}$$

and

$$Z_0(x;\omega) = r_0(x) + k_0(x) / (i\omega) \tag{6.29}$$

Here $Z_{TM}(x;\omega)$ is the TM impedance, $Z_{SC}(x;\omega)$ the impedance of the OHC cilia, $Z_{BM}(x;\omega)$ the BM impedance, and $Z_0(x;\omega)$ the impedance (with low-pass characteristic) that is involved with the pressure generation.

By eliminating $v_{TM}(x;\omega)$ from Eqs. (6.23) and (6.24) we get the following relation between $v_{BM}(x;\omega)$ and $v_{SC}(x;\omega)$:

$$v_{SC}(x;\omega) = \frac{i\beta(x;\omega)}{\eta\delta + i(\beta(x;\omega) - \sigma^2 / \beta(x;\omega))} v_{BM}(x;\omega). \tag{6.30}$$

where

$$\beta(x;\omega) = \omega / \omega_{loc}(x). \tag{6.31}$$

From Eqs. (6.25) and (6.29) with the help of Eqs. (6.6), (6.7), (6.8), (6.12) and (6.13) follows the relation

$$P_{OHC}(x;\omega) = c_0 \omega_{loc}(x)(\varepsilon\delta - i / \beta(x;\omega))v_{SC}(x;\omega) \tag{6.32}$$

so that the OHC pressure can be written as

$$P_{OHC}(x;\omega) = c_0 \omega_{loc}(x) \frac{1 + i\varepsilon\delta\beta(x;\omega)}{\delta_{SC} + i(\beta(x;\omega) - \sigma^2 / \beta(x;\omega))} v_{BM}(x;\omega), \tag{6.33}$$

with $c_0 = c_{00}m_{BM} = 0.055$, $\varepsilon\delta = 1$, $\delta_{SC} = \eta\delta = 0.14$. This activity distribution is the same as the one used in previous papers (Kanis and de Boer, 1993a, 1993b, 1994a, 1994b). The only difference is that c_0 is taken somewhat smaller to reduce instabilities occurring in the time-domain model. This results in 8 dB less amplification at the peak of the response.

The model is made nonlinear by replacing Eq. (6.5) with

$$P_{OHC}^{NL}(x,t) = P_0 \tanh(P_{OHC}(x,t) / P_0). \tag{6.34}$$

with $P_0 = 2$ [kg m^{-1} s^{-2}]. In the frequency domain the relevant transformations are given in Kanis and de Boer (1993b).

In both the time-domain and frequency-domain computations the number of sections is equal to 180, and the length of the cochlea is equal to 0.012 [m]. We made the stimulus amplitude increase as a function of time during the first five periods of the stimulus tone (in the case of two-tone suppression the first five periods of the primary component with the lower frequency). This reduces undesired low-frequency components in the model response.

6.3 Results

As a first check we compare frequency-domain solutions with time-domain responses for *single-tone* stimulation at 7 kHz.

In Fig. 1 the BM velocity response is shown as a function of location in the cochlea with the stapes located at $x = 0$. The input level of 20 dB is too small to compress the pressure generator noticeably so that the cochlea behaves nearly linearly. In this and all other figures the reference of the input level of 2 [kg m^{-1} s^{-2}] has been chosen so that 20 dB corresponds to 20 dB SPL in previous reports by us (cf. Kanis and de Boer, 1993b, Figure 2). Part (a) shows the amplitude of the response, and part (b) the phase. Dashed lines indicate the quasilinear response, and solid lines refer to the first-order component of the time-domain solution during steady state (after 4 milliseconds). The agreement between time and frequency domain solutions is excellent. The computation time of the quasilinear response was 10 seconds while the time-domain computation took more than 4 minutes on a 486 machine operating at 33 MHz.

In Fig. 2 BM velocity responses have been obtained for two different input levels, 60 and 80 dB, respectively. We see that the response is compressed for higher levels of stimulation. Furthermore, the phase lag increases (decreases) at locations basally (apically) to the peak and the phase curve exhibits a transition at the peak location. The agreement between the two method is again excellent.

The agreement is also very good in Fig. 3 where the primary component (curve 1) together with the third (curve 3) and fifth (curve 5) Fourier components are shown. The input level of the primary tone was 70 dB. This figure should be compared with Fig. B1 of Kanis and the Boer (1993b, Appendix B).

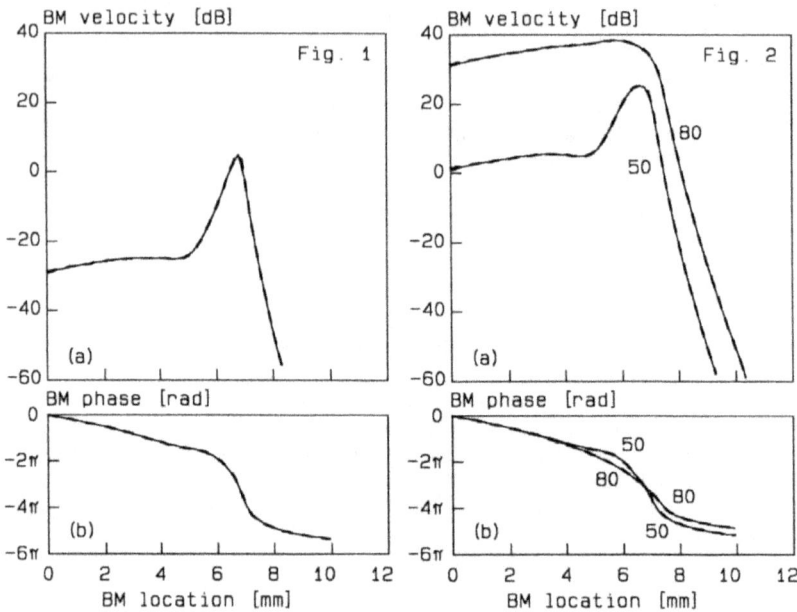

Figure 1. Single-tone stimulation at 7 kHz at an input level of 20 dB SPL. The BM velocity response is shown as a function of location in the cochlea. The stapes located at $x = 0$ [m]. Part (a) shows the amplitude of the response, and part (b) the phase. Dashed lines indicate the quasilinear response, and solid lines refer to the first-order component of the time-domain solution during steady state (which is reached after 4 milliseconds).

Figure 2. Single-tone stimulation at 7 kHz for two different input levels of 60 and 80 dB SPL, as indicated in the figure. For each quasilinear response we iterated 8 times; for the time-domain responses we needed more than 1000 iterations. See also legend to Fig. 1.

In Fig. 4 results for *two-tone* stimulation are shown. The higher tone indicated by the label 1 has a frequency of 8 kHz and an input level of 65 dB. The lower tone of 7 kHz, labelled 2, has an input level of 30 dB. The curve labelled 3 denotes the single-tone case when only the 7 kHz tone is present. We clearly see that when both tones are present the lower tone is suppressed and that the phase lag shown in part (b) of the figure has increased at locations around the peak. This figure is similar to Fig. 3 of Kanis and de Boer (1994a). It is not identical since in the above-mentioned paper we have used a slightly different model[2]. Again the two responses overlap.

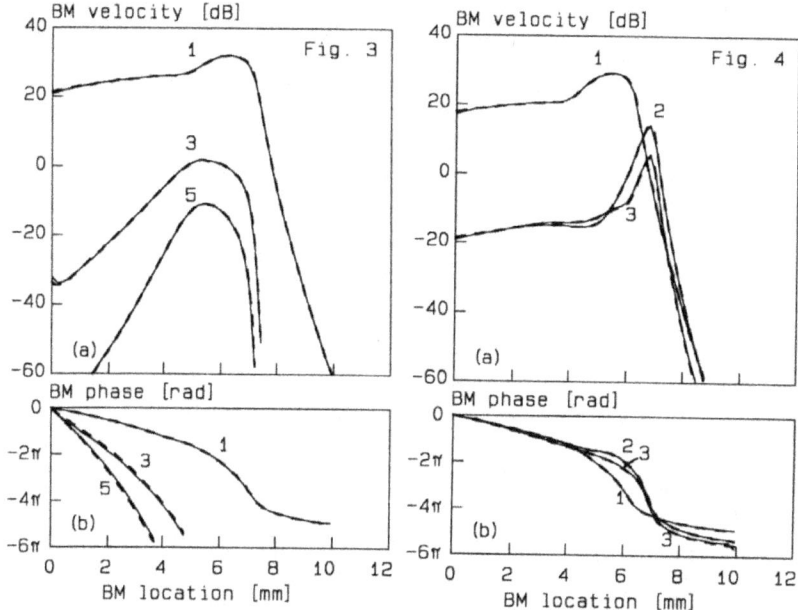

Figure 3. Single-tone stimulation at 7 kHz at an input level of 70 dB. The third-order (curve 3) and fifth-order (curve 5) distortion product are also shown. See also legend to Figure 1.

Figure 4. Two-tone suppression. The higher tone (curve 1) has a frequency of 8 kHz and an input level of 65 dB SPL. The lower tone is 7 kHz (curve 2) has an input level of 30 dB. The curve labelled 3 denotes the single-tone case. See also legend to Figure 1.

In Fig. 5 both primaries and three combination tones are shown when the cochlea is stimulated by a pair of tones with frequencies f_1 equal to 10 kHz (curve 1) and f_2 equal to 12 kHz (curve 2). Both input tones have equal levels of 60 dB. The combination tones shown are those with frequency $2f_1$-f_2 (curve 3), $3f_1$-$2f_2$ (curve 4) and $4f_1$-$3f_2$ (curve 5). We observe a difference of 1 dB between the two solutions near the peak region of the $2f_1$-f_2 combination tone. There is also a, very slight, difference between the two components with frequency $3f_1$-$2f_2$. The small differences did not decrease when we decreased the time step or increased the number of sections.

The overlap between the two solutions becomes less good near the stapes when the primary frequencies approach each other. In Fig. 6 the cochlea model has been

---

[2] In this paper we have put the $\omega_{loc}(x)$ function in Eq. (5.32) inside the nonlinearity in Eq (5.33). In Kanis and de Boer (1994a) we have put it outside the nonlinearity.

91

stimulated with two tones with frequencies of 6.25 (curve 1) and 6.5 kHz (curve 2). Input levels are 50 dB SPL. The combination tone with frequency $2f_1$-f_2 (6 kHz) has been shown as the curve 3. Again the dashed lines refer to the quasilinear case, while the solid lines correspond to the time-domain solutions. We see that a substantial difference between the two solutions exists in the region from the stapes to about $x = 5$ mm. This difference cannot be reduced by decreasing the time-step or increasing the number of sections.

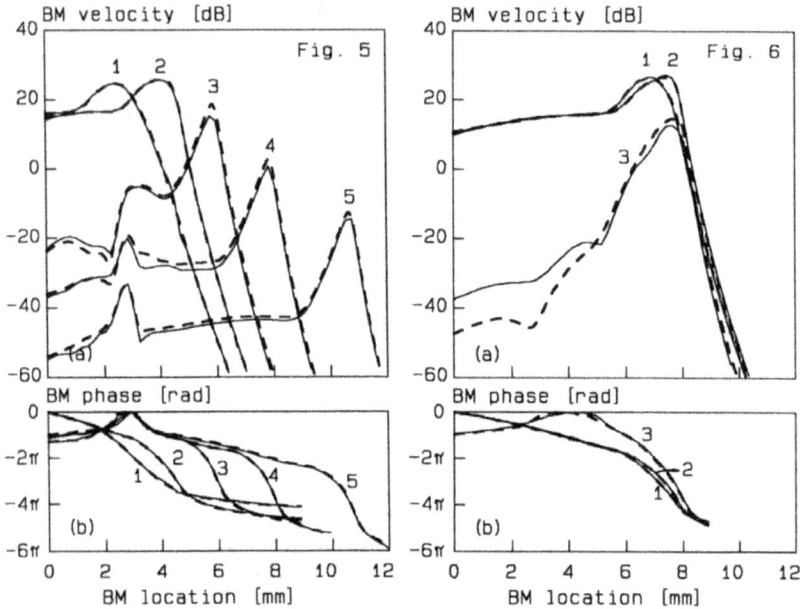

Figure 5. Combination tones. Curve 1 and 2 denote the two primaries with frequencies of 12 and 10 kHz, respectively. Both are at an input level of 60 dB SPL. Three combination tones with frequency $2f_1$-f_2 (curve 3), $3f_1$-$2f_2$ (curve 4), and $4f_1$-$3f_2$ (curve 5) are shown. Dashed lines indicate the quasilinear responses, and solid lines refer to the time-domain solution during steady state.

Figure 6. Combination tones. Primary frequencies are 6.5 (curve 1) and 6.25 (curve 2) and input levels are 60 dB SPL. Curve 3 denotes the $2f_1$-f_2 combination tone. Dashed lines indicate the quasilinear responses, and solid lines refer to the time-domain solution during steady state. Note that the two combination tone responses do not overlap at the tail of the responses.

The difference might also be created by distortion products (of higher order than that of the 6 kHz combination tone) that have been neglected by the quasilinear method. For instance, the two combination tones with frequency of 5.75 and 5.5 kHz also generate a combination tone of 6 kHz. To check whether these two combination tones influence the response of the 6 kHz combination tone in Fig. 6 we have computed, under the same stimulus conditions as for Fig. 6, the 5.75 and 5.5 kHz combination tones. We have added the responses of the two combination tones of 5.75 and 5.5 kHz to the primary responses and these summed responses were taken as input to the nonlinear pressure generator. We have then used the component in the OHC pressure at 6 kHz to compute the 6 kHz component in the velocity response. This component did not show any deviation from the response at 6 kHz in Fig. 6. Our conclusion is that we do not need to consider combination tones with higher order than the ones we are computing. The reason, of course, is that these higher-order products are much smaller than the primary components.

6.4 Discussion

In the past several nonlinear cochlea models were solved in the time-domain (Hall, 1974, 1977; Diependaal et al., 1987; Van den Raadt, 1990; Geisler et al., 1993; Neely and Stover, 1993; Cohen and Furst, 1993). A disadvantage of using the time domain is that it is difficult to understand why certain nonlinear effects in the cochlear response occur. For instance, suppressed tones behave so differently under different stimulus conditions that to some authors (Nuttall and Dolan, 1993) different mechanisms appear to be at work. To understand the nonlinear effects more clearly we developed a method to solve (stable) nonlinear models in the frequency domain. We reported on this method and its solutions in previous papers (Kanis and de Boer, 1993a, 1993b, 1994a, 1994b).

In Kanis and de Boer (1993b) we have supported the statement that the quasilinear method is valid in the case of single-tone stimulation. Further support is given in the present paper by direct comparison of the two methods, see Figs. 1 to 3 (for single-tone stimulation). In the case of two-tone stimulation we have never made a check on the accuracy of the computations. Therefore, we present comparisons in Figs. 4 and 5. Again, the difference between the primaries computed with the two methods is too small to be seen. Thus, the assumption that we do not need consider higher-order products in order to calculate the primaries (as we have done in the quasilinear case) is justified.

We have not been able to reduce the differences between the two solutions at the frequency of the combination tones. Decreasing the time step or increasing the number of sections had no effect. If we assume that the time-domain solutions are the 'correct' ones, we may conclude that for primary frequency ratios near 1.0 (as is the case in Fig. 6) the emission of the $2f_1\text{-}f_2$ combination is estimated as too small when computed with the quasilinear method. This means that the results performed in Kanis and de Boer (1994b) are influenced slightly. However, the main conclusions remain the same.

The good performance of the quasilinear method (for most situations) paves the way for more complex types of stimulation, e.g., in cases with noise and impulse signals. However, it is possible that for these multi-component stimuli the computation speed becomes smaller than that of the time-domain method. Since the quasilinear method facilitates the acquisition of insight into nonlinear processes when compared with time-domain methods, this is not a problem.

It should be noted that the nonlinearity of the active mechanism should be of the saturating type with a linear character at low inputs. A nonlinearity with expansive terms like the one used by Hall (1974, 1977) does not have the proper form for our method to be of any use (a model with this type of nonlinearity does not converge at high levels when solved with the quasilinear method).

With this proviso in mind we conclude that stable nonlinear models can be solved in the frequency domain. The quasilinear method guarantees very fast computation of the model results. The advantage with respect to the time domain becomes even larger if one considers the number of iteration steps. To compute the quasilinear responses in Fig. 6 we used 21 iteration steps in which we solved the model, whereas for the computation of the time-domain responses we solved the model more than 8000 times. With this in mind, we may understand why the use of the quasilinear method is necessary to solve nonlinear models that already take very much time to solve in the linear case. An example is the three-dimensional finite-element model by Kolston and Ashmore (1994), a linear model which reflects the anatomy of the Organ of Corti more closely than previous cochlea models. Depending on the resolution of the model, solving their model takes several hours of CPU time on an IBM-compatible personal computer with a 80486 (66 MHz) processor. Solving a nonlinear version of their model in the time-domain would take several years. In the frequency domain with use of the quasilinear method the computation time would be reduced to several days.

Acknowledgements

This work was supported by the Netherlands Foundation for Scientific Research, project number 810-410-10-1.

References

Boer, E. de (1980). "Auditory Physics. Physical principles in hearing theory. I," Phys. Rep. 62, 87-174.

Boer, E. de (1991). "Auditory Physics. Physical principles in hearing theory. III," Phys. Rep. 203, 125-231.

Diependaal, R.J., Duifhuis, H., Hoogstraten, H.W., and Viergever, M.A. (1987). "Numerical methods for solving one-dimensional cochlear models in the time domain," J. Acoust. Soc. Am. 82, 1655-1666.

Geisler, C.D., Bendre, A., and Liotopoulos, F.K. (1993). "Time-domain modeling of a nonlinear, active model of the cochlea," in *Biophysics of Hair-cell Systems*, edited by H. Duifhuis, J.W. Horst, P. van Dijk and S.M. van Netten (World Scientific, Singapore), pp. 330-337.

Hall, J.L. (1974). "Two-tone distortion products in a nonlinear model of the basilar membrane," J. Acoust. Soc. Am. 56, 1818-1828.

Hall, J.L. (1977). "Two-tone suppression in a nonlinear model of the basilar membrane," J. Acoust. Soc. Am. 61, 802-810.

Kanis, L.J., and Boer, E. de (1993a). "The emperor's new clothes: DP emissions in a locally-active nonlinear model of the cochlea," in: *Biophysics of Hair Cell Sensory Systems*, edited by H. Duifhuis, J.W. Horst, P. van Dijk, and S.M. van Netten (World Scientific, Singapore), pp. 304-311.

Kanis, L.J., and Boer, E. de (1993b). "Self-suppression in a locally active nonlinear model of the cochlea: A quasilinear approach," J. Acoust. Soc. Am. 94, 3199-3206.

Kanis, L.J., and Boer, E. de (1994a). "Two-tone suppression in a locally active nonlinear model of the cochlea," J. Acoust. Soc. Am. 96, 2156-2165.

Kanis, L.J., and Boer, E. de (1994b). "Frequency dependence of acoustic distortion products in a locally active model of the cochlea," submitted to the Journal of the Acoustical Society of America.

Kolston, P.J., and Ashmore, J.F. (1995). "Finite element micromechanical modelling of the cochlea in three dimensions," J. Acoust. Soc. Am., in press

Neely, S.T. (1985). "Mathematical modeling of cochlear mechanics," J. Acoust. Soc. Am. 78, 345-352.

Neely, S.T. and Kim, D.O. (1986). "A model for active elements in cochlear biomechanics," J. Acoust. Soc. Am. 79, 1472-1480.

Neely, S.T., and Stover, L.J. (1993). "Otoacoustic emissions from a nonlinear, active model of cochlear mechanics," in: *Biophysics of Hair Cell Sensory Systems*, edited by H. Duifhuis, J.W. Horst, P. van Dijk, and S.M. van Netten (World Scientific, Singapore), pp. 64-70.

Nuttall, A.L., and Dolan, D.F. (1993). "Response to 'Comment on "Two-tone suppression of inner hair cells and basilar membrane responses in the guinea pig," ' " J. Acoust. Soc. Am. 94, 3511-3514.

Sellick, P.M., Patuzzi, R. and Johnstone, B.M. (1982). "Measurement of basilar membrane motion in the guinea pig using the Mössbauer technique," J. Acoust. Soc. Am. 72, 131-141.

Van den Raadt, M.P.M.G., and Duifhuis, H. (1990). "A generalized Van der Pol-oscillator cochlea model," in: *Mechanics and Biophysics of Hearing*, edited by P. Dallos, C.D. Geisler, J.W. Matthews, M. Ruggero, and C.R. Steele (Springer-Verlag, Berlin), pp. 227-234.

Summary

The cochlea transforms movements of the stapes into excursions of the basilar membrane. This is done in a highly frequency-selective but also in a nonlinear way. Up to now little is known about the exact mechanism of cochlear nonlinearity. To understand these phenomena, we have devised a model of the cochlea that is solved with a new computational method. The emphasis of this thesis is not only on simulating these phenomena with the model but also on gaining insight into the mechanics of the cochlea.

In order to establish a framework in which nonlinear processes can be examined a one-dimensional cochlea model was developed that describes the response of the basilar membrane (BM). We made the model nonlinear and locally-active by including elements that generate pressures in a nonlinear way. These pressures were put directly over the BM. The active elements, supposed to be the outer hair cells (OHCs), are triggered by relative movements between the underside of the tectorial membrane and the upperside of the OHCs. These relative movements give rise to a resonance that makes the activity distribution along the BM place-frequency dependent in such a way that at places basal to the peak more energy is injected into the system than absorbed by it. As a consequence the pressure wave in the cochlea is amplified. In view of experimental results (cf. Hudspeth and Corey, 1977; Patuzzi et al., 1989) we have given the nonlinearity a saturating form.

To gain more insight into cochlear mechanics we solved the model in the frequency domain by a linearization method (which we call the quasilinear method). An important advantage over time-domain methods is that the concepts of impedance, amplitude and phase angle can be used to describe the mechanics and movements of the cochlear partition. With the BM impedance we are able to analyze and clarify the effect of nonlinearity on the phase and amplitude of the BM response. Another advantage of the quasilinear method is that the computational speed is larger than for existing methods in the time-domain. This means that we were able to solve the model on a personal computer with a good graphical interface and a debugging environment which surpasses a similar environment on a mainframe. Furthermore, the large computational speed facilitated the development of the cochlea model used in this thesis: suitable parameter values had to be found for the model both to be well-converging and to give realistic responses. This process would have taken much more time to develop with a time-domain method.

The basic assumption that led to the development of the quasilinear method is that, for stimulation with a periodic waveform, the basilar membrane oscillates also periodically with a period equal to that of the stimulus. Thus, the cochlear response can be analysed perfectly in terms of its Fourier components. The method then consists of computing the relevant Fourier components of the system variables at every location of the cochlea, and of solving the relevant Fourier components of the BM response by iteration since these components depend nonlinearly on themselves. In every iteration step a linear problem is solved. First, the model is solved with certain starting values of the 'active' impedance at every location in the cochlea. Then, the OHC pressure distribution and its primary components are computed. By dividing the primary pressure components by the corresponding velocity components we obtain a new 'active' impedance distribution with which the model is solved again to get a new estimate of the BM velocity response pattern. This is done a number of times until the model response has converged sufficiently. Finally, if desired, the combination tones can be computed by solving the model with the relevant Fourier components in the OHC-generated pressure.

In chapter 2 we have given an outline of the quasilinear solution method with results for single-tone stimulation under different levels of stimulation. It is shown that the model response becomes less frequency selective and less sensitive as input level is increased. This behaviour corresponds to that found in experiments. At levels of 40 dB SPL the influence of the nonlinearity on the response and on the impedance of the cochlear partition is largest near the peak. At higher input levels the nonlinear influence extends to more basal and apical locations until a situation is reached in which the response and BM impedance are similar to those of the passive model. In the Appendix we have estimated the effect of higher-order products on the primary response. This effect is found to be negligible so that we do not need to consider higher-order products for the computation of the primaries. Note that if we were to compute many higher-order distortion products the computation time might eventually become larger than for time-domain methods.

In chapter 3 we treat the phenomenon of two-tone suppression. This phenomenon may occur if a secondary tone is added to a single-tone stimulus: the amplitude and phase of the single-tone response may change in the presence of the second tone. In an attempt to explain this phenomenon an hypothesis was set up by Sachs and Abbas (1976). This hypothesis, called the *attenuation* hypothesis, states that attenuating the input level of a single tone has a similar effect on the response as the addition of a second tone. However, in many stimulus conditions the hypothesis has proved to be wrong. It is shown that contradictions arising from this hypothesis can be solved by considering a cochlea model like the one used here in which nonlinearity and activity are intrinsically linked. Also, by viewing the relation between the BM impedance and the phase of the BM response it can be understood why certain phase changes of the BM response occur under certain stimulus conditions. It has not been checked whether the primary solutions are affected by the presence of combination tones: that subject is deferred until chapter 6. However, it is reasonable to expect that the suppressive effect of combination tones on the primaries is small since the combination tones are small in amplitude compared to the primaries.

In chapter 4 we deal with emissions of the $2f_1$-f_2 combination tone where f_1 and f_2 are the frequencies of the primaries (with $f_2 > f_1$). Experiments have shown that otoacoustic emissions of distortion products are 'tuned' as a function of primary frequency ratio f_2/f_1. That is, if primary frequencies are changed in such a way that the frequency of the $2f_1$-f_2 combination tone remains fixed and if primary levels are kept fixed, maximal emissions occur at a primary frequency ratio of about 1.2 (this value depends on the value of the primary levels). According to several authors (Brown and Gaskill, 1990; Brown and Williams, 1993; Allen and Fahey, 1993) this 'tuning' would be due to filtering of distortion products inside the cochlea. Because the OHCs are embedded inside the Organ of Corti, the pressures generated by them can be assumed to be filtered before they are coupled back to the BM. In the current chapter we challenge the view that the experimentally observed 'tuning' is caused by this filtering. It is shown that the same degree of 'tuning' can be reached with the same cochlea model as used in the previous chapters in which there is no filtering of distortion products. The 'tuning' in the model is found to be the consequence of saturation of the active mechanism. It is not debated whether there exists a form of DP filtering (which is likely because the OHCs are embedded inside the Organ of Corti), but it is questioned whether the tuning seen in acoustic distortion data is the result of this filtering.

In chapter 5 an experiment by Allen and Fahey (1992) was replicated with our model. In this experiment they tried to disprove the existence of the cochlear amplifier by estimating the power gain in the cochlea from otoacoustic emissions of distortion products. The emission of the cubic difference tone (CDT) with frequency $2f_1$-f_2 was measured as a function of the primary frequencies f_1 and f_2 (f_2>f_1) while keeping both the CDT's frequency and the neural response of a nerve fibre tuned to this frequency constant. The CDT emission was found not to vary as primary frequencies were varied and Allen and Fahey concluded that the cochlea must be passive. This conclusion is challenged in the current chapter. Despite a maximum velocity gain of more than 40 dB in our locally-active model with respect to a passive model, we obtained essentially the same results as Allen and Fahey in their experiments: the CDT emissions changed little as primary frequencies were varied over the range used in their experiment. With this in mind, it is concluded that Allen and Fahey's interpretation of their experiment is incorrect. Careful analysis shows why they did not find any difference in emission: the range over which they varied the primary frequencies was not large enough.

In chapter 6 we have compared quasilinear responses with time-domain solutions in order to check the accuracy of the quasilinear solution method described in previous chapters. We have done this first for sinusoidal stimulation. It is shown that the overlap between model solutions obtained with the quasilinear and the time-domain method is amazing. The same may be said for the primary responses during two-tone stimulation. It is only (in the case of two-tone stimulation) for the $2f_1$-f_2 combination tone that deviations between the two methods occur. These deviations are too small to modify the conclusions from the previous chapters.

Although several nonlinear phenomena have been successfully reproduced qualitatively and quantitatively with the model used in this thesis, the model should not be conceived of as final. There are several aspects that can be improved. For instance, the model is not entirely realistic (anatomically) in that the pressure source has not been included in the Organ of Corti. (We have taken advantage of this aspect in our discussion of 'tuning' of acoustic distortion products in chapter 4.) In our model the OHCs move the BM much the way Baron von Münchhausen pulled himself out of the swamp by his own hair. By looking at the anatomy of the cochlea we find the OHCs embedded inside the Organ of Corti so that the action of the 'cochlear amplifier' must have an effect both upon the BM and on the cuticular plate. There are models that take this aspect into consideration, but they have other deficits. For instance, Neely and Stover's model (1993) does not have a 40 dB enhancement of the BM response compared to the passive model, and Geisler's model (1991) uses a stiffness value of the OHC cilia that is much larger than that measured experimentally by Strelioff and Flock (1984). Furthermore, Geisler has used a stiffness value for the tectorial membrane that is about ten times larger than that measured by Zwislocki and Cefaratti (1989). In this respect the model by de Boer (1993) is more realistic. In his model the TM consists of two stiff parts connected by a hinge so that the TM appears to be much less stiff than its constituent parts if the TM is lifted from the Organ of Corti as was done by Zwislocki and Cefaratti. Furthermore, the stiff parts of the TM couple the OHCs to the fluid in the inner spiral sulcus in such a way that a proper counterforce is provided for the OHCs (cf. a tree branch for von Münchhausen).

Another anatomical feature that might be included in our model is the longitudinal tilting of outer hair cells (Völdrich, 1983). Although several models have been devised on this basis (Kolston, 1989; Steele *et al.*, 1993; Geisler, 1994) it is not clear whether this idea will prove to be important for the future of cochlear modelling. The same applies to Kolston's (1988) idea of splitting the BM into two parts, the pars pectinata and the pars arcuata, each having different anatomical properties. We should also mention here the three-dimensional finite-element model by Kolston and Ashmore (1995) which reflects the anatomy of the Organ of Corti more closely than previous cochlea models. The importance of certain anatomical components and features can be assessed with their model, and this knowledge can be used for the development of newer, more realistic 'lumped' element cochlea.

It is not only the anatomy that is important in cochlear modelling but also its physiology. For instance, Brundin *et al.* (1989) found that isolated OHCs have tuning properties that are determined by the location of the OHCs along the BM. This finding may also have important bearing upon the development of future models.

But what about the future of the quasilinear method? In this thesis we have only used sinusoidal stimuli or combinations of these. This has been very successful. Using the method for click-stimuli is also possible, but this might turn out to be cumbersome. In the time domain solving the model will be straightforward but it still has to be examined whether, for instance, the effective BM impedance can be computed in the time domain. If that is the case, comparisons between the effective BM impedance and the BM response are possible in the time-domain the way we did in the frequency domain. Another future application of the quasilinear method might be in the implementation of a nonlinear version of the above-mentioned model by Kolston and Ashmore (1995). Since their model is computationally very slow, solving it in the time domain is not practical at present. Therefore, a faster method of computation such as the quasilinear method is called for.

Finally we would like to note that the conditions under which the long-wave approximation is valid are not fulfilled at all locations in the cochlea. In fact, they are only fulfilled at locations near the stapes. Still the one-dimensional model is a good model in that it replicates many properties of cochlear responses. In the future, computations with the quasilinear method should be done with two- or three-dimensional models. Also, we might apply the method to a transmission line model that supports both long and short waves as has been developed by Van den Raadt (personal communication). Such a model could be used to refine the simulation of nonlinear processes in the cochlea.

In this thesis we have shown that nonlinear models can be solved in the frequency domain with the quasilinear method in such a way that much insight into the mechanism of cochlear nonlinearity is obtained. Furthermore, we have shown that different cochlear phenomena can be simulated with a model in which activity and nonlinearity are elegantly combined into one mechanism. These phenomena were reproduced to a better extent than was possible with previous nonlinear models.

References

Allen, J.B. and Fahey, P.F. (1992). "Using acoustic distortion products to measure the cochlear amplifier gain on the basilar membrane," J. Acoust. Soc. Am. 96, 178-188.

Allen, J.B., and Fahey, P.F. (1993). "Evidence for a second cochlear map," in: *Biophysics of Hair Cell Sensory Systems*, edited by H. Duifhuis, J.W. Horst, P. van Dijk, and S.M. van Netten (World Scientific, Singapore), pp. 296-302.

Boer, E. de (1992). "The sulcus connection. On a mode of participation of outer hair cells in cochlear mechanics," J. Acoust. Soc. Am. 93, 2845-2859.

Brown, A.M., and Gaskill, S.A. (1990). "Measurement of acoustic distortion reveals underlying similarities between human and rodent mechanical responses," J. Acoust. Soc. Am. 88, 840-849.

Brown, A.M., and Williams, M.W. (1993). "A second filter in the cochlea," in: *Biophysics of Hair Cell Sensory Systems*, edited by H. Duifhuis, J.W. Horst, P. van Dijk, and S.M. van Netten (World Scientific, Singapore), pp. 72-77.

Brundin, L., Flock, Å, and Canlon, B. (1989). "Tuned motile responses of isolated cochlear hair cells," Acta Otolaryngol. Suppl. 467, 229-234.

Geisler, C.D. (1991). "A cochlear model using feedback from motile outer hair cells," Hear. Res. 54, 105-117.

Geisler, C.D. (1994). "A cochlear model using feed-forward outer-hair-cell forces," in preparation.

Hudspeth, A.J. and Corey, D.P. (1977). "Sensitivity, polarity, and conductance change in the response of vertebrate hair cells to controlled mechanical stimuli," Proc. Natl. Acad. Sci. USA 74, 2407-2411.

Kolston, P.J. (1988). "Sharp mechanical tuning in a micromechanical cochlear model," J. Acoust. Soc. Am. 83, 1481-1486.

Kolston, P.J., Viergever, M.A., de Boer, E., and Diependaal, R.J. (1989). "Realistic mechanical tuning in a micromechanical cochlear model, J. Acoust. Soc. Am. 86, 133-140.

Kolston, P.J., and Ashmore, J.F. (1995). "Finite element micromechanical modelling of the cochlea in three dimensions," submitted for publication in the J. Acoust. Soc. Am.

Neely, S.T., and Stover, L.J. (1993). "Otoacoustic emissions from a nonlinear, active model of cochlear mechanics," in: *Biophysics of Hair Cell Sensory Systems*, edited by H. Duifhuis, J.W. Horst, P. van Dijk, and S.M. van Netten (World Scientific, Singapore), pp. 64-70.

Patuzzi, R.B., Yates, G.K., and Johnstone, B.M. (1989). "Outer hair cell receptor current and sensorineural hearing loss," Hear. Res. 42, 47-72.

Sachs, M.B., and Abbas, P.J. (1976). "Phenomenological model for two-tone suppression," J. Acoust. Soc. Am. 60, 1157-1163.

Steele, C.R., Baker, G., Tolomeo, J., and Zetes, D. (1993). "Electro-mechanical models of the outer hair cell," in: *Biophysics of Hair Cell Sensory Systems*, edited by H. Duifhuis, J.W. Horst, P. van Dijk and S.M. van Netten (World Scientific, Singapore), pp. 207-2140-337.

Strelioff, D., and Flock, Å. (1984). "Stiffness of sensory-cell hair bundles in the isolated guinea pig cochlea," Hear. Res. 15, 19-28.

Võldrich, L. (1983). "Experimental and Topographic Morphology in Cochlear Mechanics, in: Mechanics of Hearing, edited by E. de Boer and M.A. Viergever (Martinus Nijhoff, Amsterdam), pp. 163-167.

Zwislocki, J.J, and Cefaratti, L.K. (1989). "Tectorial membrane II: Stiffness measurements *in vivo*," Hear. Res. 42, 211-227.

Samenvatting

Dit proefschrift behandelt verschillende aspekten van cochleaire niet-lineariteit. Het doel is te onderzoeken of bepaalde niet-lineaire fenomenen met elkaar gerelateerd zijn. De nadruk ligt niet zozeer op het simuleren van deze verschijnselen met een cochlea model als wel op het verkrijgen van inzicht in de mechanica van het binnenoor.

Het oor is een zeer gespecialiseerd orgaan dat geluidsignalen converteert in zenuwpulsen die auditieve informatie naar de hersenen voeren. Het oor kan in drie delen opgesplitst worden: het *uitwendige oor*, het *middenoor* en het *binnenoor* (bestaande uit de *cochlea* en de *semicirculaire kanalen*). Het uitwendige oor vangt een geluidsgolf op en leidt deze door de *gehoorgang* naar het *trommelvlies*. Dit wordt zó gedaan dat de geluidsdruk[1] ter hoogte van het trommelvlies versterkt wordt in het frekwentiegebied dat belangrijk is voor het spraakverstaan. Door de 'kleur' van het geluid te veranderen stelt het uitwendige oor ons in staat om geluid (in beperkte mate) te lokaliseren. Het middenoor bestaat uit een keten van drie gehoorbeentjes, het *aambeeld*, de *hamer* en de *stijgbeugel*. Deze keten geleidt de geluidstrillingen van het trommelvlies naar het ovale venster, de ingang van de cochlea. Het middenoor zorgt voor een 20-voudige versterking van het geluid. Deze versterking vermindert de reflektie die onstaat als het geluid van de lucht in de gehoorgang (via het middenoor) naar de vloeistof die zich in de cochlea bevindt gaat. Een andere eigenschap van het middenoor is het beveiligen van de cochlea tegen overmatig geluid. Dit mechanisme treedt niet alleen in werking als hard geluid het oor binnenkomt maar ook gedurende iemands eigen spraak (zelfs voordat die spraak begonnen is).

Als het ovale venster wordt bewogen door de stijgbeugel (op een oscillerende wijze), ontstaat er een geluidsgolf in de vloeistof van de cochlea. De geassocieerde geluidsdruk in die vloeistof doet de partitie (die de cochlea in twee delen splitst) bestaande uit het *basilair membraan* (BM) en het *orgaan van Corti* (waar de zintuigcellen zich bevinden) bewegen. Door de interactie tussen vloeistof en partitie onstaat er een lopende golf op het BM en in de vloeistof. In het geval dat de stimulus sinusvormig is stijgt de amplitude van de golf langs de lengte van de partitie tot er een maximum wordt bereikt op een plaats die afhangt van frekwentie. Hoge frekwenties vinden hun maximum op plaatsen dichtbij de stijgbeugel, terwijl lagere frekwenties op lokaties verder in de cochlea terechtkomen. De reden voor het onstaan van dit 'toonladder effect' is dat de stijfheid die toegeschreven kan worden aan de cochleaire partitie afneemt van de stijgbeugel tot aan het *helicotrema* (het andere eind van de cochlea). Als de stimulus uit meer dan één component bestaat, roept elke component een lopende golf in het leven die elk op een andere plaats zijn maximum heeft. De cochlea verricht derhalve een soort frekwentie-naar-plaats analyse van de auditieve stimulus.

In het orgaan van Corti zijn twee typen zintuigcellen aanwezig, de *binnenste* en *buitenste haarcellen*. Gedurende bewegingen van het BM onstaat er een relatieve beweging tussen de onderkant van het *tectoriaal membraan* en de toppen van deze haarcellen. Dit leidt tot uitwijkingen van de *trilhaartjes* van de haarcellen wat weer polarisatie van de cellen tot gevolg heeft. In het geval van de binnenste haarcellen leidt

---

[1] De geluidsdruk is het verschil tussen de druk veroorzaakt door de aanwezigheid van een longitudinale compressiegolf en de barometerdruk.

polarisatie tot het genereren van aktiepotentialen in de vezels van de *gehoorzenuw*. Bij de buitenste haarcellen leidt polarisatie tot veranderingen in hun eigen cellengte. Deze veranderingen zijn hoogstwaarschijnlijk betrokken bij het verbeteren van de gevoeligheid en de frekwentie selektiviteit van het oor.

Vóór 1971 dacht men dat de BM respons een lineaire funktie van het input niveau was. In 1971 liet Rhode echter zien dat voor sinusvormige stimulatie de BM respons zich niet-lineair gedraagt als een functie van input niveau: de respons wordt gecomprimeerd voor sterkere stimuli. Niet alleen de amplitude maar ook de fase van de respons wordt beïnvloed door de niet-lineariteit. Rhode en Robles (1974) en Sellick *et al.* (1982) reporteerden een groter wordende fase achterstand bij een groter wordend stimulus niveau voor frekwenties beneden de karakteristieke frekwentie (dit is de frekwentie waarvoor de bestudeerde locatie in de cochlea het meest gevoelig is). Er zijn meer niet-lineaire verschijnselen ontdekt. Gedurende twee-toon stimulatie kan de respons op een toon gesupprimeerd worden door de aanwezigheid van de andere toon. Bij dit verschijnsel, dat twee-toon suppressie wordt genoemd, vertoont de fase van de probe respons veranderingen die zeer afhangen van de stimulus condities (Cheatham en Dallos, 1990; Nuttall en Dolan, 1993). Deze kritische afhankelijkheid is in dit proefschrift onderzocht. Een ander niet-lineair verschijnsel dat ontstaat als de cochlea met twee tonen wordt gestimuleerd is de generatie van combinatie tonen. Als mensen luisteren naar toonparen kunnen ze tonen horen die niet in de stimulus aanwezig zijn (Goldstein, 1967; Smoorenburg, 1972). Pendanten van deze tonen zijn gemeten in neurale responsen (Goldstein en Kiang, 1968), in binnenste haarcellen (Nuttall en Dolan, 1990), in otoakoestische emissies (Kemp, 1979) maar ook op het niveau van bewegingen van het BM (Robles *et al.*, 1991).

Hoewel in het verleden verschillende niet-lineaire modellen van de cochlea zijn ontwikkeld, zijn de bovengenoemde niet-lineare verschijnselen nog niet geheel begrepen en gerepliceerd. Het hoofddoel van dit proefschrift is (1) het ontwikkelen van een cochlea model dat deze niet-lineaire verschijnselen beter repliceert dan voorgaande modellen en (2) het oplossen van het model op zodanige wijze dat inzicht in het werkingsmechanisme makkelijker wordt verkregen dan met bestaande oplossings-methodes.

Om een raamwerk waarin niet-lineaire processen kunnen worden onderzocht tot stand te brengen is een 1-dimensionaal cochlea model ontwikkeld dat de respons van en het drukverschil over het BM beschrijft. We hebben het model niet-lineair en lokaal aktief gemaakt door elementen te introduceren die geluidsdrukken genereren op een niet-lineaire wijze. Deze geluidsdrukken worden direkt over het BM gezet. De aktieve elementen, waarvan aangenomen wordt dat het de buitenste haarcellen (outer hair cells: OHCs) zijn, worden in werking gesteld door relatieve bewegingen tussen de onderkant van het tectoriaal membraan en de bovenkant van de OHCs. Deze relatieve beweging zorgt voor een resonantie die de aktiviteitsverdeling langs het BM plaats-frekwentie afhankelijk maakt op een dusdanige wijze dat op plaatsen basaal van de piek meer energie in het systeem wordt gepompt dan erdoor wordt geabsorbeerd. Het gevolg is dat de drukgolf in de cochlea versterkt wordt. Er is voor gezorgd dat de modelrespons goed aansluit bij experimentele resultaten verkregen bij lage stimulus niveau's. Rekening houdend met experimentele resultaten (bijv. Huspeth en Corey, 1977; Patuzzi *et al.*, 1989) hebben we de niet-lineariteit van deze OHCs een saturerende vorm gegeven.

Om meer inzicht in cochleaire mechanica te verkrijgen hebben we het door ons ontwikkelde cochlea model opgelost in het *frekwentiedomein* m.b.v. een linearisatie methode (die we de quasi-lineaire methode hebben genoemd). Een belangrijk voordeel boven tijdsdomein methodes is dat de concepten impedantie, amplitude en fasehoek kunnen worden gebruikt bij het beschrijven van de mechanica en de bewegingen van de cochleaire partitie. Met de BM impedantie kunnen we het effect van de niet-lineariteit op de fase en amplitude van de BM respons analyseren en verhelderen. Een ander voordeel van de quasilineaire methode is dat de rekensnelheid veel groter is dan voor bestaande methodes in het tijdsdomain. Dit betekent dat we het model kunnen oplossen met een PC, die een goede grafische omgeving ondersteunt. Verder is het 'ontluizen' op een PC veel handiger dan op een mainframe. De grote rekensnelheid heeft de ontwikkeling van het cochlea model dat gebruikt is in dit proefschrift vergemakkelijkt: geschikte parameter waarden moesten gevonden worden om het model zowel goed convergerend te maken als goede responsen te laten genereren. Dit proces zou veel meer tijd hebben gekost met een tijdsdomein methode.

De basis aanname die tot de ontwikkeling van de quasilineaire methode heeft geleid is dat, voor stimulatie met een periodieke golfvorm, het basilair membraan óók periodiek trilt met een periode die gelijk is aan die van de stimulus. Dit betekent dat de cochleaire respons perfect kan worden geanalyseerd in termen van zijn Fourier componenten. De methode bestaat dan ook uit het berekenen van de relevante Fourier componenten van de systeemvariabelen op elke plaats van de cochlea, en uit het oplossen van de BM respons door iteratie omdat de sterkte en fase van de Fourier componenten op een niet-lineaire wijze afhangen van de BM respons. In elke iteratiestap wordt dan een lineair probleem opgelost. Eerst wordt het model opgelost door de 'aktieve' impedantie op elke plaats in de cochlea een bepaalde startwaarden te geven. Dan worden de OHC drukverdeling en zijn primaire componenten berekend. Door de primaire drukcomponenten te delen door de corresponderende snelheidscomponenten verkrijgen we een nieuwe 'aktieve' impedantie verdeling waarmee het model opnieuw wordt opgelost om een nieuwe schatting van het patroon van de BM snelheidsrespons te bepalen. Dit wordt enkele malen gedaan totdat voldoende convergentie is opgetreden. Tot slot, indien gewenst, worden de combinatietonen berekend door het model op te lossen met de relevante vervormings-componenten in de door de OHCs gegenereerde druk.

In hoofdstuk 2 worden het in dit proefschrift gebruikte cochlea model en de quasi-lineaire oplossingsmethode beschreven. Vervolgens wordt het model opgelost met de ontwikkelde methode. Model resultaten laten een verscherping van de cochleaire frekwentie selektiviteit en een versterking van de BM respons zien als op de juiste manier drukbronnetjes (lees: buitenste haarcellen) in de cochlea worden geïntroduceerd. De plaats-frekwentie afhankelijkheid is zodanig dat in een begrensd gebiedje basaal t.o.v. de piek de weerstandscomponent van de BM impedantie een negatieve waarde heeft. Elders is de weerstand positief om stabiliteit van het model te waarborgen. Omdat elk drukbronnetje voorgesteld wordt als een saturerende funktie van zijn input, is de respons een niet-lineaire funktie van de geluidsintensiteit. Versterking van het ingangsniveau heeft een 'afvlakkende' werking op de respons: de scherpte van de respons piek en de frekwentie selektiviteit van het model nemen af. Dit gedrag wordt selfsuppressie genoemd. Bij intensiteiten boven de 80 dB SPL ziet de respons er hetzelfde uit als die van een passief model, i.e., een model zonder drukbronnetjes. De input-output funktie van de model respons is vergelijkbaar met die gemeten in experimenten (Sellick *et al.,*

105

1982; Robles *et al.*, 1986). In de appendix van dit hoofdstuk is een schatting gemaakt van de invloed van hogere-orde produkten op de primaire respons. Deze invloed blijkt nihil te zijn zodat we de hogere-orde termen niet mee hoeven te nemen in onze berekeningen.

Hoofdstuk 3 behandelt het verschijnsel twee-toon suppressie. Dit verschijnsel kan plaats vinden als een tweede toon wordt toegevoegd bij een enkele-toon stimulus: de amplitude van de enkele-toon respons kan dan kleiner worden. Tevens kan de fase veranderen. In een poging dit verschijnsel te verklaren hebben Sach en Abbas (1976) de *verzwakkings hypothese* opgesteld. Deze hypothese stelt dat verzwakking van het ingangsniveau van een enkele toon hetzelfde effect heeft op de respons van die toon als de toevoeging van een tweede toon. Echter, de hypothese heeft tot veel contradicties geleid. Deze tegenstellingen kunnen worden opgeheven met het model dat in dit proefschrift wordt gebruikt, een model waarin niet-lineariteit en aktiviteit intrinsiek verbonden zijn met elkaar. Bovendien lost het model problemen op die uit de experimenten van Nuttall en Dolan (1993) volgden. Uit hun experiment bleek dat onder bijna dezelfde omstandigheden een totaal ander fase gedrag van de gemeten respons volgde. Hun verklaring was dat er verschillende soorten suppressie zouden moeten bestaan die onder verschillende omstandigheden in werking zouden treden. Deze verklaring wordt door ons model weerlegd: al het door hun gevonden fase gedrag kan gerepliceerd en verklaard worden met het in dit proefschrift gebruikte model.

In hoofdstuk 3 behandelen we emissies van de $2f_1$-f_2 combinatietoon waarin f_1 en f_2 de frekwenties van de primaire tonen voorstellen (met $f_2 > f_1$). Experimenten hebben laten zien dat otoakoestische emissies van vervormingsprodukten 'getuned' zijn als funktie van de primaire frekwentie verhouding f_2/f_1. M.a.w., als de primaire frekwenties op zodanige manier worden veranderd dat de frekwentie van de combinatietoon gelijk blijft, dan zijn (voor gelijkgehouden primaire niveau's) de emissies maximaal bij een frekwentie verhouding van ongeveer 1.2 (deze waarde is afhankelijk van de hoogte van de primaire niveau's). Volgens verschillende auteurs (Brown en Gaskill, 1990; Brown en Williams, 1993; Allen en Fahey, 1993) is deze 'tuning' het resultaat van filtering van de vervormingsprodukten binnenin de cochlea. Omdat de buitenste haarcellen ingebed zijn in het orgaan van Corti, worden de door deze cellen gegenereerde drukken gefilterd voordat ze teruggekoppeld worden naar het basilair membraan en datzelfde geldt voor de vervormingsprodukten. In dit hoofdstuk vallen we het idee aan dat de experimenteel gevonden 'tuning' veroorzaakt wordt door zo'n filtering. We laten zien dat dezelfde graad van 'tuning' bereikt wordt in ons model waarin geen filtering van vervormingsprodukten plaatsvindt. Daarbij wordt aangetoond dat de 'tuning' het gevolg is van de in het model aanwezige niet-lineariteit. Wanneer f_2/f_1 tot 1 nadert, uit de niet-lineariteit zich gedeeltelijk in het supprimeren van het aktieve mechanisme dat de combinatietoon versterkt en gedeeltelijk in het satureren van de generatie van de combinatietoon.

In hoofdstuk 5 staat een nabootsing van een experiment door Allen en Fahey (1992) beschreven. Met het experiment probeerden ze het bestaan van de 'cochleaire versterker' te ontzenuwen. Dit deden ze door een schatting te maken van de 'power gain' in de cochlea m.b.v. otoakoestische emissies van vervormingsprodukten. De emissie van de combinatietoon (CT) met frekwentie $2f_1$-f_2 werd gemeten als functie van de primaire frekwenties f_1 and f_2 ($f_2 > f_1$) terwijl de CT's frekwentie en de neurale respons van een zenuwvezel afgestemd op deze frekwentie konstant werd gehouden. De CT emissie bleek niet te veranderen als de primaire frekwenties werden gevariëerd en Allen en Fahey konkludeerden daaruit dat de cochlea passief is. Deze konklusie wordt aangevallen in dit hoofdstuk. Ondanks een maximum versterking van de snelheidsrespons van 40 dB in een lokaal-aktief model t.o.v. een passief model, verkregen we nagenoeg dezelfde resultaten als Allen en Fahey in hun experimenten. Onze konklusie is dat Allen en Fahey's interpretatie van hun experimenten fout is. Zorgvuldige analyse toont aan waarom ze nauwelijks een verschil hebben gemeten: de range waarover ze de primaire frekwenties variëerden was niet groot genoeg om een substantieel verschil in emissie te kunnen verwachten. We hebben hun experimenten uitgebreid door bij een kleinere frekwentie verhouding te kijken de CT emissie te berekenen. Deze emissie bleek groter te zijn dan de emissie gemeten bij kleinere frekwentie verhoudingen. Toekomstig onderzoek moet uitmaken of dit ook in experimenten gevonden wordt.

In hoofdstuk 6 worden quasi-lineaire responsen met tijdsdomein oplossingen vergeleken. Het doel van deze vergelijking is om inzicht te verkrijgen in de nauwkeurigheid van de quasi-lineaire oplossingsmethode die in dit proefschrift gebruikt is. Voor sinusvormige stimulatie blijkt de overlap tussen de modeloplossingen verkregen met de quasi-lineaire en de tijdsdomein methode spectaculair goed te zijn. Voor de primaire tonen onder twee-toon stimulatie geldt hetzelfde. Alleen voor de $2f_1$-f_2 combinatietoon wijken de responsen verkregen met de twee methodes in lichte mate af. Deze afwijkingen zijn te klein om gevaar op te leveren voor de konklusies gemaakt in dit proefschrift.

In dit proefschrift hebben we laten zien dat niet-lineaire modellen in het frekwentie domein kunnen worden opgelost met een quasi-lineaire oplossingsmethode op een zodanige wijze dat veel inzicht wordt verkregen in het werkingsmechanisme van cochleaire niet-lineariteit. Verder is aangetoond dat verschillende cochleaire verschijnselen kunnen worden gesimuleerd met een model waarin aktiviteit en niet-lineariteit tot één geheel is verweven. Deze fenomenen zijn met meer succes gesimuleerd dan mogelijk is met vroegere modellen, en zijn daardoor beter te begrijpen.

Referenties

Allen, J.B. and Fahey, P.F. (1992). "Using acoustic distortion products to measure the cochlear amplifier gain on the basilar membrane," J. Acoust. Soc. Am. 96, 178-188.

Allen, J.B., and Fahey, P.F. (1993). "Evidence for a second cochlear map," in: *Biophysics of Hair Cell Sensory Systems*, edited by H. Duifhuis, J.W. Horst, P. van Dijk, and S.M. van Netten (World Scientific, Singapore), pp. 296-302.

Brown, A.M., and Gaskill, S.A. (1990). "Measurement of acoustic distortion reveals underlying similarities between human and rodent mechanical responses," J. Acoust. Soc. Am. 88, 840-849.

Brown, A.M., and Williams, M.W. (1993). "A second filter in the cochlea," in: *Biophysics of Hair Cell Sensory Systems*, edited by H. Duifhuis, J.W. Horst, P. van Dijk, and S.M. van Netten (World Scientific, Singapore), pp. 72-77.

Cheatham, M.A., and Dallos, P. (1990). "Comparison of low- and highside two-tone suppression in inner hair cell and organ of Corti responses," Hear. Res. 50, 193-210.

Goldstein, J.L. (1967). "Auditory nonlinearity," J. Acoust. Soc. Am. 41, 676-689.

Goldstein, J.L. and Kiang, N.Y.S. (1968). "Neural correlates of the aural combination tone 2f1-f2," Proc. IEEE 56, 981-992.

Hudspeth, A.J. and Corey, D.P. (1977). "Sensitivity, polarity, and conductance change in the response of vertebrate hair cells to controlled mechanical stimuli," Proc. Natl. Acad. Sci. USA 74, 2407-2411.

Kemp, D.T. (1979). "Evidence of mechanical nonlinearity and frequency selective wave amplification in the cochlea," Arch. Otorhinolaryngol. 224, 37-45.

Nuttall, A.L., and Dolan, D.F. (1990). "Inner hair cell responses to the $2f_1$-f_2 intermodulation distortion product," J. Acoust. Soc. Am. 87, 782-790.

Nuttall, A.L., and Dolan, D.F. (1993). "Response to 'Comment on "Two-tone suppression of inner hair cells and basilar membrane responses in the guinea pig," ' " J. Acoust. Soc. Am. 94, 3511-3514.

Patuzzi, R.B., Yates, G.K., and Johnstone, B.M. (1989). "Outer hair cell receptor current and sensorineural hearing loss," Hear. Res. 42, 47-72.

Rhode, W.S. (1971). "Observations of the vibration of the basilar membrane in squirrel monkeys using the Mössbauer technique," J. Acoust. Soc. Am. 49, 1218-1231.

Rhode, W.S., and Robles, L. (1974). "Evidence from Mössbauer experiments for nonlinear vibration in the cochlea," J. Acoust. Soc. Am. 55, 588-596.

Robles, L., Ruggero, M.A. and Rich, N.C. (1986). "Basilar membrane mechanics at the base of the chinchilla cochlea. I. Input-output functions, tuning curves, and response phases," J. Acoust. Soc. Am. 80, 1364-1374.

Robles, L., Ruggero, M.A. and Rich, N.C. (1991). "Two-tone distortion in the basilar membrane of the cochlea," Nature 349, 413-414.

Sachs, M.B., and Abbas, P.J. (1976). "Phenomenological model for two-tone suppression," J. Acoust. Soc. Am. 60, 1157-1163.

Sellick, P.M., Patuzzi, R. and Johnstone, B.M. (1982). "Measurement of basilar membrane motion in the guinea pig using the Mössbauer technique," J. Acoust. Soc. Am. 72, 131-141.

Smoorenburg, G.F. (1972). "Combination tones and their origin," J. Acoust. Soc. Am. 52, 615-632.

Publications and poster presentations

Kanis, L.J. and Boer, E. de (1991). "Suppression and self-suppression in a quasi-linear active model of the cochlea," poster presented during the International Meeting on *Auditory Processing of Complex Sounds* at the Royal Society in London, England.

Breed, A.J., Kanis, L.J. and Boer, E. de (1992). "Cochlear nonlinearity for complex stimuli," in: *Auditory physiology and perception*, edited by Y. Cazals, L. Demany and K. Horner (Pergamon Press, Oxford), pp. 189-195.

Kanis, L.J. and Boer, E. de (1992). "Combination tones and emissions in an active nonlinear model of the cochlea," poster presented during the 3rd International Symposium on *Cochlear Mechanisms and Otoacoustic Emissions* in Rome, Italy.

Kanis, L.J., and Boer, E. de (1993). "The emperor's new clothes: DP emissions in a locally-active nonlinear model of the cochlea," in: *Biophysics of Hair Cell Sensory Systems*, edited by H. Duifhuis, J.W. Horst, P. van Dijk, and S.M. van Netten (World Scientific, Singapore), pp. 304-311.

Kanis, L.J., and Boer, E. de (1993). "Self-suppression in a locally active nonlinear model of the cochlea: A quasilinear approach," J. Acoust. Soc. Am. 94, 3199-3206.

Kanis, L.J., and Boer, E. de (1994). "Two-tone suppression in a locally active nonlinear model of the cochlea," J. Acoust. Soc. Am. 96, 2156-2165.

Kanis, L.J., and Boer, E. de (1994). "Frequency dependence of acoustic distortion products in a locally active model of the cochlea," submitted to the Journal of the Acoustical Society of America.

Kanis, L.J., and Boer, E. de (1995). "Comparison between time-domain and frequency-domain solutions for a locally-active nonlinear model of the cochlea," submitted to the Journal of the Acoustical Society of America.

Kanis, L.J., and Boer, E. de (1995). "Tuning of acoustic distortion products in a locally active cochlea model," poster presentation at the 1995 Midwinter Meeting of the Association for Research in Otolaryngology, St. Petersburg, USA.

Acknowledgements

I would like to thank:
- my promotor Prof. Dr. Egbert de Boer for his great support throughout my research. He spent large amounts of time improving my writing and stimulating my thinking about cochlearity. I enjoyed our discussions very much;
- the many people I had discussions with on Internet;
- Peter-Paul Boermans, René van der Horst, Helene de Bruijn and Sidonne van Kreveld-Bos for their company during lunch-breaks.
- Maureen Peeck-O'Toole for checking the English in parts of the manuscript.

CURRICULUM VITAE

The author of this thesis was born in Amsterdam on May 23, 1963. He attended primary school in Apeldoorn and went to the Gemeentelijk Gymnasium in the same city, matriculating in Dutch, English, Latin, Mathematics, Geometry, Physics, Biology, and Chemistry. He studied Theoretical Physics at the Rijksuniversiteit Groningen. The subject of his Master's thesis was the Gel'fand-Levitan-Marchenko equation. This equation is an inverse Schrödinger equation from which one can deduce, given a set of scattering data, the form of the potential that gives rise to these scattering data. For this project he spent six months under the supervision of Prof. Dr. F. Calogero and Prof. Dr. A. Degasperis at the University 'La Sapienza' in Rome. He continued work on the project under the supervision of Prof. Dr. D. Atkinson at the Rijksuniversiteit Groningen and obtained his Master's degree in August 1988. In January 1990 he started to work as a Ph.D. student on a project called "The influence of nonlinearity on the frequency-selectivity of the ear". This research was performed at the Academic Medical Centre in Amsterdam under the supervision of Prof. Dr. Egbert de Boer who is one of the world's foremost researchers on cochlear mechanics. Results of this research were presented at international conferences and in the form of the papers which constitute the present dissertation.

www.ingramcontent.com/pod-product-compliance
Lightning Source LLC
Chambersburg PA
CBHW071226170526
45165CB00003B/1007